Psychiatric Diagnosis

Psychiatric Diagnosis

Fourth Edition

DONALD W. GOODWIN, M.D.
Professor and Chairman
Department of Psychiatry
University of Kansas School of Medicine

SAMUEL B. GUZE, M.D.
Spencer T. Olin Professor of Psychiatry
and Head of the Department
Vice Chancellor for Medical Affairs
Washington University School of Medicine

New York / Oxford
OXFORD UNIVERSITY PRESS
1989

Oxford University Press

Oxford   New York   Toronto
Delhi   Bombay   Calcutta   Madras   Karachi
Petaling Jaya   Singapore   Hong Kong   Tokyo
Nairobi   Dar es Salaam   Cape Town
Melbourne   Auckland

and associated companies in
Berlin   Ibadan

Published by Oxford University Press, Inc.,
200 Madison Avenue, New York, New York 10016

Oxford is a registered trademark of Oxford University Press

Library of Congress Cataloging-in-Publication Data

Goodwin, Donald W.
  Psychiatric diagnosis.

  Includes bibliographies and index.
    1. Mental illness—Classification.   2. Mental
illness—Diagnosis.   I. Guze, Samuel B., 1923–
II. Title.   [DNLM: 1. Mental Disorders.   2. Mental
Disorders—diagnosis.   WM 100 G656p]
RC455.2.C4G66   1989   616.89'075   88–9972
ISBN 0–19–505230–7
ISBN 0–19–505231–5 (pbk.)

Acknowledgment is hereby made for permission to reprint
material from American Psychiatric Association, *Diagnostic
and Statistical Manual of Mental Disorders*, Third Edition,
Revised (Washington, D.C.: APA, 1987).

9 8 7 6 5 4 3 2 1

Printed in the United States of America
on acid-free paper

To Eli Robins and the late Robert A. Woodruff, Jr.

Preface to the Fourth Edition

"The paradox of modern psychiatry is that while its practice remains uncertain, the scientific problems it poses are more interesting and exciting than those in any other branch of medicine."

—Editorial in *Lancet*, October 17, 1987

Sixteen years have passed since this book was first published. In that time an interesting change has occurred in American psychiatry. Psychiatrists have become diagnosis-conscious.

DSM-III, the official diagnostic manual of the American Psychiatric Association, was published in 1980 and became an immediate best-seller. Previous diagnostic and statistical manuals had caused hardly a ripple. Why the change?

Possibly fatigued by unsupported theory, psychiatrists may have found some satisfaction in agreeing on what to call things. If illnesses could not be explained—and doubters were everywhere—at least they could be described. Also, with the discovery of relatively specific drug therapies, diagnosis had become *practical*. With the availability of lithium and neuroleptic drugs, distinguishing between mania and schizophrenia—once an interesting academic exercise—might now determine how a patient was treated.

With the emphasis on diagnosis, use of words like *data, reliability,* and *operational* increased. Journals brimmed over with graphs and tables; circles with arrows pointing this way and that appeared less and less. People actually stood up at meetings and asked: "Where is the evidence?"

These were real changes. We would like to think *Psychiatric Diagnosis* anticipated them. The first edition contained diagnostic criteria similar to the criteria later adopted by DSM-III. Its philosophy was described as "agnostic"; DSM-III calls its approach "atheoretical." The first edition presented the features of a dozen disorders; DSM-III described 226—a difference in quantity, not kind.

That *Psychiatric Diagnosis* was a forerunner of DSM-III should not be surprising. The authors were faculty members at Washington University in St. Louis. A third of the DSM-III task force consisted of Washington University–trained psychiatrists. In the sixties and seventies Washington University's Department of Psychiatry gained the reputation of having a strong alliance with biology, taking a medical view of psychiatric illness and placing heavy reliance on reproducible data. It believed in diagnosis, and the belief caught on.

A revised edition of DSM-III was published in 1987 (DSM-III-R). Apart from the criteria for substance abuse, changes were relatively minor. Some new categories were added. One was Late Luteal Phase Dysphoric Disorder, hormonalese for premenstrual syndrome. Some women object to premenstrual syndrome being called a psychiatric disorder. Changing the name was intended to appease them but didn't. They picketed APA headquarters and held long sessions with lawyers. The term got in anyway, but in a Needing Further Study section.

The fourth edition of *Psychiatric Diagnosis* has added no new categories. It takes the position that only about a dozen diagnostic entities in adult psychiatry have been sufficiently studied to be useful. (For more on this heretical view, see the Preface to the First Edition.) The two DSM-IIIs introduced many new diagnostic terms. The terms used in this book were chosen because, in our judgment, they are the ones most widely used in psychiatry. Where they differ from those recommended by DSM-III-R, the reason for using the older, better-known, term is provided in the "Definition" section of each chapter.

The fourth edition contains much new information. This is reflected in the number of references. There are now 1,056 references, three times greater than in the first edition. Each reference refers to something in the text: if readers want to know where we got that idea, we usually can tell them. (The idea still may be wrong, of course.) The practice of identifying sources in a textbook has never, to our knowledge, been applied so exhaustively.

This new edition also differs from previous editions in other significant ways. Treatment sections have been expanded. Clinical vignettes have been added. Recent findings on biological aspects of psychiatric illness have been incorporated. Lastly, a chapter on "The Psychiatric Examination" has been added.

In general, however, the book hews to its original goal: to provide a concise compendium of current knowledge in psychiatry, with abundant citations, not much theory, and as little personal opinion as we could get by with.

November 1988   D.W.G.
                                                         Kansas City
                                                           S.B.G.
                                                          St. Louis

Preface to the First Edition

Because it remains a rose.

Classification has two functions: communication and prediction. A rose can be defined precisely. It has pinnate leaves, belongs to the rose family, and so forth. When you say "rose" to a person who knows something about the definition, communication results.

A rose also has a predictable life history: it stays a rose. If it changes into a chrysanthemum, it may not have been a rose in the first place. If roses routinely change into chrysanthemums, like caterpillars into butterflies, well and good. Natural history may include metamorphoses but they must be routine to be "natural."

Classification in medicine is called diagnosis, and this book is about diagnosis of psychiatric conditions. Diagnostic categories— diseases, illnesses, syndromes—are included if they have been sufficiently studied to be useful. Like roses, they can be defined explicitly and have a more or less predictable course.

In choosing these categories, the guiding rule was: *diagnosis is prognosis*. There are many diagnostic categories in psychiatry, but few are based on a clinical literature where the conditions are defined by explicit criteria and follow-up studies provide a guide to prognosis. Lacking these features, such categories resemble what sociologists call labeling. Two examples are "passive-aggressive personality" and "emotionally unstable personality," which, like most personality diagnoses, have been inadequately studied for us to know whether they are useful or not.

Not every patient can be diagnosed by using the categories in this book. For them, "undiagnosed" is, we feel, more appropriate than a label incorrectly implying more knowledge than exists. Terms like "functional" and "psychogenic" and "situational reaction" are sometimes invoked by physicians to explain the unexplained. They usually mean "I don't know," and we try to avoid them.

Because classification in psychiatry is still at a primitive stage, there are reasonable grounds for questioning our choice of categories. In general, we lump rather than split. Hence we have two affective disorders—primary and secondary—rather than the half-dozen affective disorders cited in the official nomenclature. Schizophrenia is divided into "good prognosis" and "bad prognosis" schizophrenia rather than sliced more finely, as some prefer. Our justification for this is "the literature," meaning primarily follow-up studies.

"The follow-up is the great exposer of truth, the rock on which many fine theories are wrecked and upon which better ones can be built," wrote P. D. Scott. "It is to the psychiatrist what the postmortem is to the physician." Not all such studies are perfect, but we feel they are better than no studies. And inevitably there are instances where our "clinical judgment" has prevailed in evaluating the merit of individual studies. No text in psychiatry could be written today without a certain amount of this, but we have tried to limit personal opinion to a minimum. Many if not most assertions have a citation, and the reader can check the references to form his own judgment.

When the term "disease" is used, this is what is meant: a disease is a cluster of symptoms and/or signs with a more or less predictable course. Symptoms are what patients tell you; signs are what you see. The cluster may be associated with physical abnormality or may not. The essential point is that it results in consultation with a physician who specializes in recognizing, preventing, and, sometimes, curing diseases.

It is hard for many people to think of psychiatric problems as diseases. For one thing, psychiatric problems usually consist of symptoms—complaints about thoughts and feelings—or behavior disturbing to others. Rarely are there signs—a fever, a rash. Almost never are there laboratory tests to confirm the diagnosis. What people say changes from time to time, as does behavior. It is usually harder to agree about symptoms than about signs. But whatever the psychiatric problems are, they have this in common with "real" diseases—they result in consultation with a physician and are associated with pain, suffering, disability, and death.

Another objection to the disease or medical "model" arises from a misconception about disease. Disease often is equated with physical abnormality. In fact, a disease is a category used by physicians, as "apples" is a category used by grocers. It is a useful category if precise and if the encompassed phenomena are stable over time. Diseases are conventions and may not "fit" anything in nature at all. Through the centuries, diseases have come and gone, some more useful than others, and there is no guarantee that our present diseases"—medical or psychiatric—will represent the same clusters of symptoms and signs a hundred years from now that they do today. On the contrary, as more is learned, more useful clusters surely will emerge.

There are few explanations in this book. This is because for most psychiatric conditions there *are* no explanations. "Etiology unknown" is the hallmark of psychiatry as well as its bane. Historically, once etiology is known, a disease stops being "psychiatric." Vitamins were discovered, whereupon vitamin-deficiency psychiatric disorders no longer were treated by psychiatrists. The spirochete was found, then penicillin, and neurosyphilis, once a major psychiatric disorder, became one more infection treated by nonpsychiatrists.

Little, however, is really known about most medical illnesses. Even infectious diseases remain puzzles in that some infected individuals have symptoms and others do not.

People continue to speculate about etiology, of course, and this is good if it produces testable hypotheses, and bad if speculation is mistaken for truth. In this book, speculation largely is avoided, since it is available so plentifully elsewhere.

A final word about this approach to psychiatry. It is sometimes called "organic." This is misleading. A better term, perhaps, is agnostic. Without evidence, we do not believe pills are better than words. Without evidence, we do not believe chemistry is more important than upbringing. Without evidence, we withhold judgment.

Advocacy is not the purpose of this book. Rather, we hope it will be useful in applying current knowledge to those vexatious problems—crudely defined and poorly understood—that come within the jurisdiction of psychiatry.

St. Louis   D.W.G.
December 1973

Contents

Psychiatric Diagnosis

1. Affective Disorders

Definition

. . . There is a pitch of unhappiness so great that the goods of nature may be entirely forgotten, and all sentiment of their existence vanish from the mental field. For this extremity of pessimism to be reached, something more is needed than observation of life and reflection upon death. The individual must in his own person become the prey of pathological melancholy. . . . Such sensitiveness and susceptibility to mental pain is a rare occurrence where the nervous constitution is entirely normal; one seldom finds it in a healthy subject even where he is the victim of the most atrocious cruelties of outward fortune . . . it is positive and active anguish, a sort of psychical neuralgia wholly unknown to healthy life.

WILLIAM JAMES (54)

Depression and euphoria are the primary symptoms of affective disorders, but not the only ones. Associated with low moods are such symptoms as insomnia, anorexia, suicidal thoughts, and feelings of worthlessness or of being a burden to others; associated with euphoria are such symptoms as hyperactivity and flight of ideas. The extent of depression or euphoria is often inappropriate to the patient's life situation, a fact sometimes as obvious to patients as to their relatives and friends.

DSM-III-R has renamed affective disorders, calling them mood disorders. Mood is defined as a "prolonged emotion that colors the whole psychic life . . . generally involving depression or elation." When there is a choice between newer and older terms, this book generally opts for those most clinicians use, which in this case is the older term "affective disorders." Hence the title of the chapter remains the same as in previous editions.

Whatever the name, the definition of affective (or mood) disorders remains unchanged. It refers to a group of disorders characterized by prolonged disturbances of mood, "accompanied by a full or partial manic or depressive syndrome . . . not due to any other physical or mental disorder" (DSM-III-R).

Affective disorders have been divided and subdivided endlessly as investigators endeavor to distinguish "normal" from "abnormal" mood and to create clinical clusters with distinct natural histories, familial prevalence, course and prognoses, and response to treatment. After a century there is still no agreement about the most satisfactory classification.

Amid all this diversity, there is a common theme: Affective disorders are primarily characterized by depressed mood, elevated mood (mania), or alternations of depressed and elevated moods. The classical term for the latter is manic depressive illness; a newer term is bipolar disorder. The two are interchangeable.

Milder forms of a depressive syndrome are called dysthymic disorder; milder forms of mania are called hypomania; and milder expressions of bipolar disorder are called cyclothymic disorder. DSM-III-R criteria for each are presented in Tables 1.1–1.5.

Confounding the nosologic confusion is the use of two other terms: "primary affective disorder" and "secondary affective disorder." As used in this book primary affective disorder refers to individuals who have had no previous psychiatric disorder or else only episodes of depression or mania. Secondary affective disorder refers to patients with a preexisting psychiatric illness other than depression or mania. Affective disorder is defined as bipolar when mania occurs, whether depressions occur or not. When the disorder involves solely depression, the term "unipolar" affective disorder is often used.

For all these terms, more than disturbed mood is required. There must be a syndrome, a group of characteristic clinical features that distinguish one disorder from another.

Historical Background

Descriptions of affective disorder began with Hippocrates. The term "melancholia" is usually attributed to him, as is the notion that it results from the influence of black bile and phlegm on the brain "darkening the spirit and making it melancholy . . ." (66).

About five hundred years later, early in the second century A.D., Aretaeus of Cappadocia recognized and recorded an association between melancholia and mania: "Those affected with melancholia are not everyone of them affected according to one particular form; they are either suspicious of poisoning or flee to the desert from misanthropy, or turn superstitious, or contract a hatred of life. If at any time a relaxation takes place, in most cases hilarity supervenes . . . the patients are dull or stern, dejected or unreasonably torpid, without any manifest cause . . . they also become peevish, dispirited, sleepless, and start up from a disturbed sleep. Unreasonable fear also seizes them, if the disease tends to increase . . . they complain of life, and desire to die" (59). Aretaeus observed that affective disorder was often episodic but also occurred in a chronic, unremitting form. Like Hippocrates, he attributed the cause to a humoral imbalance: "If it [black bile] be determined upwards to the stomach and diaphragm, it forms melancholy, for it produces flatulence and eructations of the fetid and fishy nature, and it sends rumbling wind downwards and disturbs the understanding" (66).

The nineteenth-century French physician Falret described an episodic variety of depression with remissions and attacks of increasing duration, an illness occurring more frequently among women than men, sometimes associated with precipitating events, sometimes alternating with mania (la folie circulaire). Falret and his contemporary Baillarger (who also described recurring attacks of mania and melancholia) probably influenced Kraepelin's later concept of manic depressive psychosis.

In 1896 Kraepelin made his major contribution to psychiatry by separating the functional psychoses into two groups, dementia praecox and manic depressive psychosis. Dementia praecox was chronic and unremitting with a generally bad prognosis. Manic depressive psychosis, on the other hand, did not usually end in chronic invalidism. After publishing the sixth edition of his textbook in 1896, Kraepelin continued to define the limits of dementia praecox narrowly, but he expanded those of manic depressive psychosis to include almost all abnormalities of mood. Patients with chronic depressions were included as well as those with episodic illness, manics as well as depressives (62, 124).

Kraepelin had insisted that manic depressive psychosis was generally independent of social and psychological forces, that the cause of the illness was "innate." Freud and the psychoanalysts assumed the opposite. Freud in *Mourning and Melancholia*, published in 1917, outlined his theories of the psychodynamic genesis of depression (40). He hypothesized that depression had in common with the process of mourning a response to the loss of a "love-object," that is, the loss of something greatly valued. Grief, a healthy response, differed from melancholia in that the latter involved *intense* expression of ambivalent, hostile feelings, formerly associated with the object. Upon loss of the loved person or thing, these unresolved, negative feelings were directed inward, resulting in despair, a sense of worthlessness, thoughts of self-harm, and other depressive symptoms.

Since the second decade of the twentieth century there has been considerable controvery over the distinction between "endogenous" depression and "reactive" depression. This controversy had its origin partially in the differing viewpoints of the Kraepelinians and Freudians toward mental phenomena in general. Kraepelin and his followers searched for the limits of pathological behavior by describing the symptoms of syndromes in keeping with the traditions of nineteenth-century German medicine. Freud and his pupils searched for mental mechanisms that might be most obvious in pathological states but were not limited to those states. Such differences in attitude were augmented by the fact that Kraepelinian psychiatrists dealt chiefly with severely ill, hospitalized patients, whereas Freudian psychiatrists tended to treat mildly ill, nonhospitalized patients. The differences have never been fully resolved. There has been classification after classification of the affective disorders, most frequently in terms of dichotomies: endogenous opposed to reactive, psychotic opposed to neurotic, agitated opposed to retarded.

A large part of the twentieth-century literature on affective disorders assumes tacitly that two basic forms of depression do exist. The terms "reactive" and "neurotic" have often been equated, as have the terms "endogenous" and "psychotic." It has been assumed that depressions of the former type are milder, that they are more often a direct result of precipitating events or unique individual responses to social and psychological stress, and more recently that they are less responsive to somatic therapy. However, controversy persists about the validity of this distinction. An alternative that avoids inference about cause is the classification of affective disorders as *primary* or *secondary* (see Definition).

Reactive and endogenous depressions may be classified as either primary or secondary affective disorders. Patients with mild or severe depressions, with or without "psychotic" symptoms such as hallucinations or delusions, with many episodes or with few, and regardless of age of onset, may be diagnosed at having either primary or secondary affective disorder.

Reports have been published in the last decade indicating that primary affective disorder should be divided into bipolar and unipolar forms (21, 46, 79, 124). (One's initial impression of the terms "bipolar" and "unipolar" is that bipolar patients have had both mania and depression, whereas unipolar patients have had either one or the other alone. "Bipolar" actually refers to patients with mania whether or not they have also had depressions. "Unipolar" refers to patients who have had depression alone.) Patients with bipolar illness have a somewhat earlier age of onset than unipolar patients. Their histories are characterized by more frequent and shorter episodes, even when depression is considered alone. There is a greater prevalence of affective disorder among relatives of bipolar patients than among relatives of unipolar patients.

Many of the questions about affective disorders that have plagued investigators are still unresolved. At present there is no way to evaluate the importance of precipitating events in either bipolar or unipolar illness. There have been preliminary efforts to divide unipolar primary affective disorder into early onset and late onset forms (125), but as yet no generally accepted subclassification of unipolar illness exists. An abiding problem is the old question of how to separate the experience of bereavement from that of depression.

Epidemiology

Estimates of the prevalence of primary affective disorder depend on the sample or population studied and on the definition of the illness. Older studies of large groups of patients selected from isolated areas—Iceland (50) and the Danish islands of Bornholm (39) and Samsø (108)—indicated that 5 percent of men and 9 per cent of women can expect to have primary affective disorder at some time during their lives. An ongoing study by the National Institute of Mental Health (NIMH) tends to confirm these findings, showing a prevalence of bipolar and unipolar disorders of about 5 percent in the general population (78). Much higher rates, however, have been reported in other studies (119); thus it is difficult to know which data are most reliable.

Problems in ascertainment may not be the only explanation for the variability. Cultural and genetic differences may be important in determining the actual prevalence of affective disorders. For example, a 1983 study of 12,500 Amish individuals in Pennsylvania yielded a prevalence rate of 1 percent for primary affective disorder (35). The Amish are a culturally and genetically homogeneous population consisting of large extended families who live in a socially cohesive environment in which alcoholism, drug abuse, and sociopathy are virtually absent. It is conceivable that reports of higher rates of affective disorder were based on samples "contaminated" by individuals with other psychiatric disorders or possibly that the Amish have a smaller genetic propensity for affective disorders than other populations. Finally, as the Amish study was one of the first to use DSM-III criteria, a third explanation is that previous studies used looser criteria that produced inflated rates because of false positives.

In any case, private psychiatric hospitals report that primary affective disorder is the problem for which patients are most frequently admitted, and the same is true of many psychiatric clinics.

Furthermore, whatever the chief diagnosis, depression is a common reason for psychiatric consultation. Among patients with panic disorder in one study, for example, half presented with a secondary depression (127). Secondary depression is also frequently the reason for psychiatric consultation among alcoholics, hysterics, and other patients who come to a psychiatric clinic (49, 127, 128). Depression—primary and secondary—is clearly the most common diagnosis in psychiatry (51, 118).

Most studies show that primary affective disorder is more common in women than men (86, 117). However, this may only apply to unipolar disorder where women, in some studies, outnumber men by two to one (117). Apparently, men and women share about equally the risk of suffering bipolar disorder (9).

Studies of hospitalized patients early in the twentieth century indicated that 30 to 50 percent of patients with manic depressive psychosis had either mania or a history of previous mania. The fact that bipolar illness is so disruptive, however, increases the likelihood of hospitalization. Experience with both inpatients and outpatients suggests that approximately one in ten affectively disordered patients is bipolar, but even this may be an overestimate of the prevalence of bipolar illnesses.

A link between bipolar affective disorder and above-average occupational or educational achievement has been suggested (126). Some studies raise the possibility that primary affective disorder especially affects socially striving individuals who feel great need for social success and approval (13, 28, 41, 50). The low prevalence of affective disorder among the Amish (35) may reflect their simple life style and unconcern for material success—although this explanation is just as speculative as the others.

Clinical Picture

Diagnostic criteria for depressive and manic episodes are presented in Tables 1.1 and 1.2 DSM-III-R also has criteria for two other affective disorders: Cyclothymic Disorder and Dysthymic Disorder (Tables 1.3 and 1.4). The latter is roughly synonymous with what previously was called "depressive neurosis." Melancholia—perhaps the oldest term of all for depression—was retained by DSM-III-R to refer to particularly severe forms of depression believed to be especially responsive to drug and electroconvulsive therapy (ECT) (Table 1.5). Criteria for hypomania are encompassed in the manic episode syndrome (Table 1.2).

Table 1.1 Diagnostic criteria for Major Depressive Episode (DSM-III-R)

A. At least five of the following symptoms have been present during the same two-week period and represent a change from previous functioning; at least one of the symptoms is either (1) depressed mood, or (2) loss of interest or pleasure. (Do not include symptoms that are clearly due to a physical condition, mood-incongruent delusions or hallucinations, incoherence, or marked loosening of associations.)

(1) depressed mood (or can be irritable mood in children and adolescents) most of the day, nearly every day, as indicated either by subjective account or observation by others

(2) markedly diminished interest or pleasure in all, or almost all, activities most of the day, nearly every day (as indicated either by subjective account or observation by others of apathy most of the time)

(3) significant weight loss or weight gain when not dieting (e.g., more than 5% of body weight in a month), or decrease or increase in appetite nearly every day (in children, consider failure to make expected weight gains)

(4) insomnia or hypersomnia nearly every day.

(5) psychomotor agitation or retardation nearly every day (observable by others, not merely subjective feelings of restlessness or being slowed down)

(6) fatigue or loss of energy nearly every day

(7) feelings of worthlessness or excessive or inappropriate guilt (which may be delusional) nearly every day (not merely self-reproach or guilt about being sick)

(8) diminished ability to think or concentrate, or indecisiveness, nearly every day (either by subjective account or as observed by others)

(9) recurrent thoughts of death (not just fear of dying), recur-

rent suicidal ideation without a specific plan, or a suicide attempt or a specific plan for committing suicide

B.　(1)　It cannot be established that an organic factor initiated and maintained the disturbance

　　(2)　The disturbance is not a normal reaction to the death of a loved one

C.　At no time during the disturbance have there been delusions or hallucinations for as long as two weeks in the absence of prominent mood symptoms (i.e., before the mood symptoms developed or after they have remitted).

D.　Not superimposed on Schizophrenia, Schizophreniform Disorder, Delusional Disorder, or Psychotic Disorder Not Other Specified.

Table 1.2 Diagnostic criteria for Manic Episode (DSM-III-R)

Note: A "Manic Syndrome" is defined as including criteria A, B, and C below. A "Hypomanic Syndrome" is defined as including criteria A and B, but not C, i.e., no marked impairment.

A.　A distinct period of abnormally and persistently elevated, expansive, or irritable mood.

B.　During the period of mood disturbance, at least three of the following symptoms have persisted (four if the mood is only irritatable) and have been present to a significant degree:

　　(1)　inflated self-esteem or grandiosity

　　(2)　decreased need for sleep, e.g., feels rested after only three hours of sleep

　　(3)　more talkative than usual or pressure to keep talking

　　(4)　flight of ideas or subjective experience that thoughts are racing

　　(5)　distractibility, i.e., attention too easily drawn to unimportant or irrelevant external stimuli

　　(6)　increase in goal-directed activity (either socially, at work or school, or sexually) or psychomotor agitation

　　(7)　excessive involvement in pleasurable activities which have a high potential for painful consequences, e.g., the person engages in unrestrained buying sprees, sexual indiscretions, or foolish business investments

C.　Mood disturbance sufficiently severe to cause marked impairment in occupational functioning or in usual social activities or relationships with others, or to necessitate hospitalization to prevent harm to self or others.

D. At no time during the disturbance have there been delusions or hallucinations for as long as two weeks in the absence of prominent mood symptoms (i.e., before the mood symptoms developed or after they have remitted).

E. Not superimposed on Schizophrenia, Schizophreniform Disorder, Delusional Disorder, or Psychotic Disorder, Not Other Specified.

F. It cannot be established that an organic factor initiated and maintained the disturbance.

Note: Somatic antidepressant treatment (e.g., drugs, ECT) that apparently precipitates a mood disturbance should not be considered an etiologic organic factor.

Table 1.3 Diagnostic criteria for Cyclothymic Disorder (DSM-III-R)

A. For at least two years (one year for children and adolescents), presence of numerous Hypomanic Episodes (all of the criteria for a Manic Episode except criterion C that indicates marked impairment) and numerous periods with depressed mood or loss of interest or pleasure that did not meet criterion A of Major Depressive Episode.

B. During a two-year period (one year in children and adolescents) of the disturbance, never without hypomanic or depressive symptoms for more than two months at a time.

C. No clear evidence of a Major Depressive Episode or Manic Episode during the first two years of the disturbance (or one year in children and adolescents).

D. Not superimposed on a chronic psychotic disorder, such as Schizophrenia.

E. It cannot be established that an organic factor initiated and maintained the disturbance, e.g., repeated intoxication from drugs or alcohol.

Table 1.4 Diagnostic criteria for Dysthymic Disorder (DSM-III-R)

A. Depressed mood (or can be irritable mood in children and adolescents) for most of the day, more days than not, as indicated either by subjective account or observation by others, for at least two years (one year for children and adolescents)

B. Presence, while depressed, of at least two of the following:

    (1) poor appetite or overeating
    (2) insomnia or hypersomnia
    (3) low energy or fatigue
    (4) low self-esteem
    (5) poor concentration or difficulty making decisions
    (6) feelings of hopelessness

C. During a two-year period (one-year for children and adolescents) of the disturbance, never without the symptoms in A for more than two months at a time.

D. No evidence of an unequivocal Major Depressive Episode during the first two years (one year for children and adolescents) of the disturbance.

Note: There may have been a previous Major Depressive Episode, provided there was a full remission (no significant signs or symptoms for six months) before development of the Dysthymia. In addition, after these two years (one year in children or adolescents) of Dysthymia, there may be superimposed episodes of Major Depression, in which case both diagnoses are given.

E. Has never had a Manic Episode or an unequivocal Hypomanic Episode.

F. Not superimposed on a chronic psychotic disorder such as Schizophrenia.

G. It cannot be established that an organic factor initiated and maintained the disturbance, e.g., prolonged administration of an antihypertensive medication.

Table 1.5 Diagnostic criteria for Melancholia (DSM-III-R)

The presence of at least five of the following:

    (1) loss of interest or pleasure in all, or almost all, activities
    (2) lack of reactivity to usually pleasurable stimuli (does not feel much better, even temporarily, when something good happens)
    (3) depression regularly worse in the morning
    (4) early morning awakening (at least two hours before usual time of awakening)
    (5) psychomotor retardation or agitation (not merely subjective complaints)
    (6) significant anorexia or weight loss (e.g., more than 5% of body weight in a month)

(7) no significant personality disturbance before first Major Depressive Episode

(8) one or more previous Major Depressive Episodes followed by complete, or nearly complete, recovery

(9) previous good response to specific and adequate somatic antidepressant therapy, e.g., tricyclics, ECT, MAOI, lithium

---

As noted earlier, the classification of affective disorders remains controversial. DSM-III, for example, does not distinguish between "primary" and "secondary" affective disorder. We make the distinction for the following reason: Although the symptoms of primary and secondary affective disorder are similar, the two conditions have different prognostic and therapeutic implications (49, 94, 125). Primary affective disorder, which occurs in the absence of a preexisting psychiatric disorder or a chronic debilitating medical illness, consists of discrete episodes interspersed with periods of normality. In the case of secondary affective disorders, when the preexisting illness is chronic, which is usually the case, the patient is *not* well between episodes. Depressions that are indistinguishable symptomatically from primary affective disorder occur commonly in obsessive compulsive disorder, phobic disorder, panic disorder, somatization disorder, alcoholism, drug dependence, and antisocial personality. In fact, almost all psychiatric disorders, including schizophrenia and brain syndromes, are associated with an increased risk of secondary depression (secondary mania is much less common).

A further advantage of the primary-secondary distinction is that some studies indicate that primary affective disorder involves a higher risk of suicide than does a secondary affective disorder (except for alcoholism, which also involves a considerable suicide risk [48]).

Finally, decisions about treatment will be influenced by distinguishing primary from secondary affective disorder; the preexisting illness as well as the depressive syndrome must be treated in the latter.

The chief complaints of patients with a depressive episode are usually psychological: feelings of worthlessness, despair, or ideas of self-harm. But it is also common for depressed patients to complain chiefly of pains, tachycardia, breathing difficulty, gastrointestinal dysfunction, headache, or other somatic disturbances (23).

The dysphoric mood experienced by patients with depressive illness is usually characterized as sadness or despondency, but some patients describe themselves as feeling hopeless, irritable, fearful, worried, or simply discouraged. Occasionally, patients will present with what seems to be primary affective disorder, though they report minimal feelings of dysphoria. Such patients may complain of insomnia and anorexia. They may even cry profusely while telling the examining physician that they do not feel sad. These patients are unusual, but not unknown to psychiatrists.

Other characteristic symptoms of depression are anorexia with weight loss; insomnia; early morning awakening; loss of energy, described as general tiredness or fatigability; agitation or (its opposite) psychomotor retardation; loss of interest in usual activities, including loss of interest in sex; feelings of self-reproach or guilt, which may be delusional in intensity; inability to focus one's thoughts, often with a simultaneous awareness of slowed thinking; and recurrent thought of death or suicide.

It is common for patients with primary affective disorder to say, "Something is wrong with my mind." Patients will often tell their physician that they fear they are losing their mind or have a sense of emotions out of control. It is also common for depressed patients to have a low expectancy of recovery. Such a pessimistic outlook should serve as a warning that the patient may be depressed. Medically ill patients seldom give up all hope of improvement, even if seriously ill.

In some depressed patients agitation is so overwhelming that other symptoms go almost unnoticed. These patients are brought to physicians when they are found by relatives or friends pacing, wringing their hands, bemoaning their fate, clinging to anyone who will listen. They ask for reassurance, they beg for help, yet nothing satisfies them.

In other patients retardation is prominent. Marked slowing of both thought and motor behavior occurs. Tasks that once took minutes may require hours. These patients may be so slowed that it is painful to listen to their conversation. Psychomotor retardation can be so severe that a patient becomes mute or even stuporous.

Paranoid symptoms can occur among patients with primary affective disorder. These are usually exaggerated ideas of reference associated with notions of worthlessness. Characteristic delusions of patients with depression are those of a hypochondriacal or nihilistic type. Some severely ill depressives seem to feel they are so guilty and evil that they have become the focus of universal abhorrence or even that the world is disintegrating because of their terrible inadequacies and failures.

Hallucinations may also occur in primary affective disorder. These commonly involve accusatory voices or visions of deceased relatives associated with feelings of guilt. Delusions and hallucinations occurring in primary affective disorder are usually "mood congruent," their content consistent with the person's dominant mood. If a depressed person has delusions and hallucinations, the themes are commonly guilt, disease, poverty, death, or deserved punishment. If a manic has delusions or hallucinations, they are often of inflated worth, power, or special relationship to a deity or famous person.

DSM-III-R includes mood-congruence as a diagnostic feature of affective disorders. Mood-incongruent psychotic features are more often seen in schizophrenia where, for example, the patient may seem cheerful and relaxed while describing terrifying delusional experiences. The importance of mood-congruence is based mainly on a widespread clinical impression rather than systematic studies.

Depressed patients may or may not mention events that they consider important in producing their illness. When a precipitating event is described, it is sometimes surprisingly trivial, difficult for the examining physician to take seriously. Furthermore, critical evaluation of the chronology of a depressed patient's symptoms may reveal that some symptoms actually began *before* the so-called precipitating event (23). This suggests that some patients who begin to feel depressed search for reasons to explain their depression, unable or unwilling to believe they could feel as they do for no apparent reason.

The stressful events women most frequently mention as precipitants of depression are pregnancy and childbirth. In one series, 37 percent of female bipolars and 17 percent of female unipolars had their first episode of depression during pregnancy or postpartum (55).

A change in drinking habits often accompanies depressive illness (23). Middle-aged individuals who begin to drink heavily and do not have a previous history of alcoholism may be suffering from depression. On the other hand, some individuals drink less than usual when depressed.

A physician (86) has written movingly about how it feels to have the illness:

Firstly, it is very unpleasant: depressive illness is probably more unpleasant than any disease except rabies. There is constant mental pain and often psychogenic physical pain too. If one tries to get such a patient to titrate other pains against the pain of his depression one tends to end up with a description that would raise eyebrows even in a medieval torture chamber.

Naturally, many of these patients commit suicide. They may not hope to get to heaven but they know they are leaving hell. Secondly, the patient is isolated from family and friends, because the depression itself reduces his affection for others and he may well have ideas that he is unworthy of their love or even that his friendship may harm them. Thirdly, he is rejected by others because they cannot stand the sight of his suffering.

There is a limit to sympathy. Even psychiatrists have protective mechanisms for dealing with such cases: the consultant may refer the patient to an outpatient clinic; he may allow too brief a consultation to elicit the extent of the patient's suffering; he may, on the grounds that the depression has not responded to treatment, alter his diagnosis to one of personality disorder—comforting, because of the strange but widespread belief that patients with personality disorders do not suffer.

Fourthly, and finally, the patient tends to do a great cover-up. Because of his outward depression he is socially unacceptable, and because of his inward depression he feels even more socially unacceptable than he really is. He does not, therefore, tell others how bad he feels.

Most depressives, even severe ones, can cope with routine work—initiative and leadership are what they lack. Nevertheless, many of them can continue working, functioning at a fairly low level, and their deficiencies are often covered up by colleagues. Provided some minimal degree of social and vocational functioning is present, the world leaves the depressive alone and he battles on for the sake of his god or his children, or for some reason which makes his personal torment preferable to death.

The cardinal features of mania are euphoria, hyperactivity, and flight of ideas. Not all manic patients are euphoric; some are irritable instead. Flight of ideas is a rapid digression from one idea to another. One's response to a manic patient is often that of sympathetic amusement. In fact, experienced clinicians who find themselves amused by a patient immediately consider the possibility that the patient is either manic or hypomanic (mildly manic). Flight of ideas, unlike the incoherence and tangentiality of schizophrenia, is usually understandable, even though some connections between ideas may be tenuous. (Comedians often use a well-controlled flight of ideas to amuse audiences.) Attention is often called to this symptom by a push of speech, that is, speech in which a great deal is said in a short period of time. Such speech may be accompanied by rhyming, punning, and jocular associations.

Psychotic symptoms also occur: persecutory and grandiose delusions, hallucinations, ideas of reference. They are usually mood-congruent. Some patients exhibit depression and mania simultaneously. They may cry while speaking euphorically or show other unusual combinations of symptoms.

The separation of affective states into unipolar and bipolar subtypes was first proposed by Leonhard et al. (65), and support for the separation has emerged from numerous studies in Europe and the United States. Compared to unipolars, bipolars tend to have an earlier age of onset (late twenties), cyclical depressions, a higher frequency of postpartum onsets, and a greater tendency for suicide attempts. Unipolar patients have a mean age of onset in the mid-thirties, tend to have single episodes, and their depressions are often confounded with anxiety (3, 33). A few studies indicate that bipolars who have had depressions respond better to lithium than unipolars (33).

Among other reported differences between bipolar and unipolar disorder, bipolars are more likely to be delusional and to manifest psychomotor retardation. Unipolar patients more often have agitated depressions (14, 20, 81).

Neurochemical correlates of bipolar and unipolar affective disorder also have been reported. Schildkraut et al. (103) devised a mathematical equation that differentiates bipolar from chronic unipolar depressions; and follow-up studies by Shildkraut et al. tend to confirm the test's validity (104). The equation utilizes data on urinary catecholamines and metabolites. Italian investigators (101) reported a correlation between a urinary metabolite of norepinephrine and outcome with lithium: Bipolar patients with low excretion of the metabolite during depression responded best to lithium, supporting the theory that low levels of norepinephrine are associated with bipolar depression. The most recent evidence that bipolar and unipolar affective disorders represent two different disorders derives from positron emission tomography (PET) studies that show different group metabolism rates for discreet cerebral structures in the two disorders (106). Bipolar patients showed a reduction of glucose metabolism in structures above the tentorium, whereas unipolar depressed patients had higher values.

Some children have episodes of depression that resemble depression in adulthood: crying, social withdrawal, hypersensitivity, and behavioral problems (84). It is not clear whether such episodes are an early manifestation of primary affective disorder. Two observations suggest that they may not be: (a) as a rule, childhood depression is more common in boys, adult depression more common in women (117, 130); (b) the sleep of depressed children does not differ from that of age-matched normal children, whereas depressed adults differ from normal subjects in taking longer to fall asleep, awakening more often throughout the night, and experiencing greater early-morning wakefullness (64).

Depression in adults is also associated with a reduction in slow-wave sleep (stages three and four), shortened rapid eye movement

(REM) latency, and increased REM density. Sleep studies of depressed children demonstrate none of these differences (130). This suggests either that a sleep disorder makes a rather abrupt appearance within the depressive syndrome or that the depressive disorders of children and adults differ in important ways. Because human growth hormone (HGH) is excreted mostly during sleep, sleep has an additional function in children not present in adults. As a physiological state, sleep may be more protected in children than in adults. In any event, it appears that sleep measures are less useful as markers for childhood depression than for depressive illness in later years.

The dexamethasone suppression test (DST) has been called psychiatry's first laboratory test. Introduced by B. J. Carroll, it has generated both enthusiasm and criticism (22). It is only fairly *sensitive* for affective disorder: About 50 percent of patients with depression test positive, that is, resist suppression of blood cortisol levels following a dose of dexamethasone. The *specificity* of the test is higher. Only about 10 percent of *normal* control subjects are nonsuppressors. However, the specificity drops to less than 70 percent in some psychiatric conditions—including dementia and alcohol abuse—and in medical conditions that involve weight loss.

These findings were reported by an American Psychiatric Association task force (113). The report stated that positive initial DST status in depression does not add to the likelihood of antidepressant response and that a negative test is not an indication for withholding antidepressant treatment. "Although the clinical utility of the DST is limited," concluded the task force, "in certain situations its thoughtful use may aid clinical decision-making."

It has been suggested that the DST might be used to determine when antidepressant treatment should be discontinued (47). Following symptom remission in depression, there is believed to be a period of increased risk of symptom recurrence. Because there is no way to determine the length of this period, it is common practice to continue medication for at least six months after the remission of acute symptoms. Normalization of previously abnormal DST cortisol levels may indicate sufficiently diminished risk of symptom recurrence so that one can safely discontinue medication. Unfortunately, evidence that normalization of the DST predicts continued remission is mixed. Pending further evidence, most clinicians continue antidepressant drugs after remission of symptoms for several months regardless of DST results.

Another laboratory diagnostic test for depression that has been proposed involves administering thyroid-releasing hormone (TRH) to suspected depressed patients. Results vary, but in some studies half or more of unipolar patients have a "blunted" response of thyroid-stimulating hormone (TSH) secretion following a challenge with TRH. Some believe that a combination of the DST and TRH tests will identify "most" paitents with unipolar depression (42), but this is debated. Even advocates of the laboratory tests do not consider them diagnostic in themselves.

Natural History

The natural history of primary affective disorder is variable. The age of risk extends through life. Together with the usual episodic nature of the illness, this distinguishes primary affective disorder from most other psychiatric illnesses.

The mean age of onset of primary affective disorder is approximately forty. For bipolar patients, the mean age of onset is about thirty (34). Several studies have indicated a significant correlation between family members in age of onset (55).

How often do patients experience a single episode of affective disorder without recurrence? At a recent conference sponsored by the NIMH (76), authorities reviewed the literature and reached these conclusions: (a) most patients who have a manic episode have multiple recurrences of depression and mania; (b) about 70 percent of patients who have a major depressive episode will have at least one subsequent episode of depression in their lifetime and about 12 percent will have subsequent manic episodes; (c) about 20 percent of patients with affective disorders develop a chronic illness difficult for physicians to manage.

Between episodes of illness, patients with primary affective dis-

order usually function well, though there are occasional residual
symptoms. Bipolars may be an exception. Winokur and colleagues
(123) studied remitted manic depressives and found that only
61 percent of the untreated subjects were symptom-free between
acute episodes. Maintained on lithium, however, bipolars between
episodes were found to be as healthy as a comparison group of
nonpsychiatric patients and their episodes were much diminished
in frequency and intensity.

The length of individual episodes is extremely varible, ranging
from a few days to many years. In one study, 50 percent of patients recovered within one year and annual rates of recovery then
declined steadily to 28 percent in the second year, 22 percent in
the third year, and 18 percent in the fourth year (57).

The period for highest risk of relapse is shortly after recovery.
About 25 percent of patients relapse within twelve weeks after recovery; 12 percent relapse within four weeks (57, 58). The rate of
relapse declines steadily thereafter. The longer the patient has
remained well, the lower is his or her current risk of relapse. Increased risk of relapse is associated with (a) the presence of a
chronic depression of at least two years' duration and (b) a history of at least three previous affective episodes (58). Treatment
differences may account for different rates of relapse. Both lithium
carbonate and antidepressant drugs prevent relapse in some patients but not others.

Complications

There is a clear association between primary depression and suicide: Of those who commit suicide 50 to 70 percent can be
found retrospectively to have had symptoms characteristic of depression. Of those who have depression, based on a review of
seventeen studies, 15 percent will eventually die by suicide (48).

Although most people who commit suicide have made previous
suicide attempts, only a small percentage of attempters eventually
kill themselves. Nine follow-up studies of attempters showed that
no more than 2 to 4 percent are dead by suicide within five years

after an attempt (83). The more medically serious the attempt, the more likely it is that a person will subsequently die by suicide—although this is a weak predictor at best—and medically trivial attempts are also sometimes followed by completed suicide.

The risk of suicide is not necessarily correlated with the severity of symptoms. An increased risk of suicide is associated with age greater than forty, with being male, and with communication of suicidal intent (95). Folklore that patients who talk of suicide do not commit suicide is untrue. Suicidal communication must be interpreted in light of the diagnosis. The two disorders most frequently associated with suicide are primary affective disorder and alcoholism (95).

Disregarding suicide, patients with primary affective disorder may still have an increased mortality compared with matched members of the general population. In two studies, it was found that depressed men had an increased mortality from physical disease, especially carcinoma (18, 60).

Alcoholism may be a complication of primary affective disorder. This is particularly true when a person begins to drink heavily in mid or late life because "primary" alcoholism usually begins earlier. Drug abuse may also be a complication of primary affective disorder.

Poor judgment is another complication of primary affective disorder. Manic patients often show poor judgment, going on spending sprees and making impulsive, unrealistic decisions. Depressed patients also make bad decisions. Decisions to leave a job, to move to a different city, or to separate from a spouse are sometimes the result of the restless dissatisfaction associated with depression. Many clinicians advise depressed patients not to make major life decisions until they are clearly in remission.

Studies of psychiatric illness in the postpartum period indicate that bipolar women are more likely to have episodes of depression or mania during the puerperium than at other times in their lives. Having had a postpartum episode of depression, the likelihood that a woman with bipolar illness will have another episode after subsequent pregnancy is high. Female relatives of bipolar women, if they have been affectively ill, are more likely to have had episodes of illness during the puerperium than at other times (5, 10, 19, 57, 91).

There is a significant relationship between unipolar affective disorder and poor academic performance, including college dropout (79, 122). In one study at a large university, students with a high number of affective symptoms received poorer grades than did others (122). Depressed students were likely to drop courses or leave college.

On psychological testing, depressed patients may show impairment in short-term but not long-term memory (109). As the clinical state improves, so does short-term memory. Depressed patients also do poorly on tests of abstract thinking, making more errors than controls on the Halstead-Reitan Category Test (102).

Sometimes the memory impairment with depression is so profound that a mistaken diagnosis of dementia is made. Memory impairment owing to depression is called "pseudodementia." Its true identity is revealed when memory returns to normal after recovery from the depression.

Family Studies

Several studies in the 1960s reached similar conclusions about familial patterns of affective disorders: Affective disorders tend to be familial and can be subdivided into bipolar and unipolar types. In general, bipolar patients have bipolar relatives and unipolar patients have unipolar relatives (9, 81, 123).

Recently, doubt has been cast on the validity of these reports. In a large collaborative study by the NIMH, relatives of bipolar patients had higher rates of bipolar illness than the relatives of unipolar patients, but the relatives of unipolar patients did *not* have a higher rate of unipolar illness than the relatives of bipolar patients (7). The study included a category called "bipolar II," which classified patients who had clear-cut depressions but hypomanic rather than manic episodes. The bipolar II syndrome (Table 1.2) tended to run in families, with 8 percent of the patients having bipolar II disorder, but only 1 percent of the relatives of bipolar II patients had bipolar I disorder, suggesting that bipolar I and II disorders were genetically distinct conditions. As the authors of this collaborative study indicate, their findings have led to a good deal of confusion regarding familial patterns in affective disorders. They express the hope that "more complex genetic analysis" may clarify some of these issues.

Meanwhile, data from adoption and twin studies provide strong evidence for the role of genetic factors in affective disorders generally (106). An adoption study showed that children of parents with affective disorder, when adopted out to foster parents without affective disorder, continued to have an elevated risk of affective disorder (74).

There have been nine twin studies (4) of affective disorder, the first conducted in 1928. Only three dealt with unipolar affective disorder as a separate group. Of a total of 83 bipolar monozygotic probands, 72 percent had a bipolar cotwin. This contrasted with a 14 percent concordance rate in 226 dizygotic twins. The concordance rate for unipolar affective disorder was 40 percent in monozygotic twins and 11 percent in dizygotic twins. Thus, bipolar illness appears to be more strongly influenced by genetic factors than unipolar illness.

An association between unipolar affective disorder and alcoholism has been proposed by Winokur and his coworkers (115). In a series of publications they have contrasted "pure depressive illness" with "depressive spectrum disease." The latter refers to families in which there is a high prevalence of alcoholism and sociopathy among the men and a high prevalence of early-onset unipolar affective disorder among the women. A study of daughters of alcoholics raised by adoptive parents compared to daughters of alcoholics raised by their own alcoholic parents revealed that the latter group did, indeed, have a high prevalence of depression but that daughters raised by adoptive parents had no more depression than did controls. This suggested that the association of depression with alcoholism in family members may be strongly influenced by environmental factors (45).

One study (77) suggests that when a patient with primary affective disorder or alcoholism has a family history of suicide attempt, the risk of suicide is increased. A family history of attempted suicide does *not* seem to predict successful suicide in patients with sociopathy, hysteria, or opiate addiction.

The mode of genetic transmission in affective disorder is nonmendelian and almost certainly polygenic. An HLA "genetic marker" for affective disorders has been reported by two groups (71,121). Both found that pairs of siblings, when both had affective disorders, shared HLA haplotypes more often than would be expected by chance. The HLA-linked genes were on chromosome 6. Two other groups (43, 111) challenged the above reports on statistical grounds and also failed to replicate the findings.

Studying a large pedigree of Amish subjects, Egland et al. (36) identified a dominant gene conferring a "strong predisposition" to manic depressive disease on the short arm of chromosome 11. However, Hodgkinson et al. (52) found no evidence for chromosomal linkage in three family pedigrees. They concluded that linkage in manic depression is heterogenous and that mutations might explain the findings in the Amish sample.

Differential Diagnosis

Making the distinction between grief and primary affective disorder can be difficult. However, grief usually does not last as long as an episode of primary affective disorder (24, 25, 26). The majority of bereaved persons experience fewer symptoms than do patients with primary affective disorder. Furthermore, some symptoms common in primary affective disorder are relatively rare among persons experiencing bereavement, notably fear of losing one's mind and thoughts of self-harm (25, 26). If bereavement persists for more than one year, it is likely that the individual has had a preexisting psychiatric disorder.

Differential diagnosis between anxiety neurosis and primary affective disorder can be difficult because anxiety symptoms occur frequently in primary affective disorder and depressive symptoms occur frequently among anxiety neurotics. The distinction depends chiefly on chronology. If anxiety symptoms antedate the depressive symptoms, the diagnosis is anxiety neurosis. If depressive symptoms appeared first, the diagnosis is primary affective disorder. Anxiety neurosis almost always begins relatively early in life. One should be cautious about diagnosing anxiety neurosis with secondary affective disorder when the illness begins in mid or late life (127).

Patients with primary affective disorder often report somatic symptoms; hysterics often report affective symptoms. If depressive and anxiety symptoms predominate, the diagnosis of hysteria should be made with caution, particularly if the illness did not occur until the patient was thirty or older. Though patients with primary affective disorder may report many somatic symptoms, these symptoms are seldom spread throughout the system review. Furthermore, conversion symptoms (unexplained neurological symptoms) combined with sexual and menstrual symptoms are infrequent in primary affective disorder.

Obsessions occur commonly in primary affective disorder. The distinction between obsessional illness and primary affective disorder is also made on the basis of chronology. If obsessions and compulsions antedate depressive symptoms, a diagnosis of primary affective disorder should not be made.

Distinguishing between schizophrenia and primary affective disorder is usually not a problem. Schizophrenia, a chronic illness of insidious onset, is not characterized by the remitting course found in primary affective disorder. Patients with primary affective disorder do not develop the formal thought disorder characteristically seen in schizophrenia. Occasionally, the distinction between mania and schizophrenia may be difficult. Bizarre and dramatic hallucinations, delusions, and other abnormalities of mental content like those seen in schizophrenia may occur in mania. A previous history of episodic illness with remission or the presence of euphoria, hyperactivity, or flight of ideas indicates that the diagnosis may be mania rather than schizophrenia. The physician should also be careful in making a diagnosis of schizophrenia when a pa-

tient has a family history of affective disorder. This history by itself favors the diagnosis of primary affective disorder (114).

For many years, clinicians assumed that schizophreniform illness (schizoaffective psychosis) was more closely related to schizophrenia than to primary affective disorder. Now there is increasing
evidence that its relation to primary affective disorder is stronger
(24, 27, 29). Family-history studies indicate that among the relatives of schizophreniform patients there is an increased prevalence
of affective disorder. After remission, patients with schizophreniform illness may become ill again in episodic fashion, as do patients with primary affective disorder. Further episodes of illness
may be typical of depression or mania (24).

A systematic study of eighty-eight psychiatric inpatients who
were diagnosed as having either manic disorder or schizoaffective
disorder revealed no differences between the two groups with regard to the clinical picture, demographic variables, individual or
family history, and treatment response (2).

Organic brain syndromes are sometimes accompanied by depressive symptoms. Among patients with chronic brain syndromes,
those with cerebral arteriosclerosis may be particularly likely to
have episodes of secondary affective disorder (97). Pseudodementia (gross memory impairment in severely depressed patients) may
lead to diagnostic confusion in deciding which is primary—the depression or the dementia (109).

Depressive or manic symptoms are side effects of certain drugs.
Reserpine or alphamethyldopa, used in the treatment of hypertension, may produce depression. Steroids may produce schizophreniform illness. Some women respond to birth control pills
with mild to moderate depressive symptoms (69).

Clinical Management

The management of depression in primary affective disorder always involves supportive psychotherapy. Many clinicians believe
that *insight-directed* phychotherapy, involving examination of mo-
tives and deep feelings, is probably not wise because it tends to increase the patient's feelings of guilt. Some recent evidence, how-

ever, indicates that certain types of psychotherapy may be useful for mild or moderate depressions. The studies aré of three types: (a) psychotherapy alone compared with control group, (b) psychotherapy compared with antidepressant drugs, and (c) combined psychotherapy and drugs compared with psychotherapy alone and drugs alone. Five types of psychotherapy have been employed in the studies: (a) cognitive therapy, (b) behavioral therapy, (c) interpersonal therapy, (d) group therapy, and (e) marital therapy.

The goal of *cognitive therapy* is to "break down existing negative cognitions and replace them with more positive functionally adaptive ones" (12). In *behavior therapy*, the patient is "trained to function effectively in interpersonal interactions so as to maximize reinforcement obtained from others" (38). "Social skills training" is one type of behavior therapy. It emphasizes assertiveness, verbal skills, and practice at making others feel more comfortable (12). *Interpersonal psychotherapy* attempts to improve the quality of the patient's social and interpersonal functioning by "enhancing ability to cope with internally and externally induced stresses, by restoring morale, and by helping the patient deal with the personal and social consequences of the disorder" (120). In *group therapy*, a psychotherapist and a group of patients attempt to effect changes in the emotional states and behavior of the patients. *Marital therapy* may be done with an individual, a couple or family, or a group of couples.

Weissman (120) reviewed seventeen clinical trials that tested the efficacy of the five psychological treatments alone in comparison, and in combination, with drug therapy in samples of depressed oupatients. In all the studies, psychotherapy was more efficacious than nontreatment. The studies that compared drugs with psychotherapy were equivocal. One study found psychotherapy more efficacious than drugs; one study found both about equal; and three studies found that drugs produce relief of symptoms, whereas psychotherapy has more effect on social functioning. Four studies examined the combination of tricyclic antidepressants and psychotherapy. All found the combination more efficacious than either treatment alone or no treatment (17, 31, 98).

Based on these studies, drugs and psychotherapy combined

seemed to be the treatment of choice for at least mild to moderate depressions. As Weissman (120) points out, many patients will not tolerate tricyclic antidepressants. Others do not wish to enter into psychotherapy. "Under either of these conditions," Weissman says, "the patient should not be denied treatment but should be offered the alternative." There are not sufficient data to say which of the psychotherapies should be used, which patients will recover quickly with no treatment, or which subtype of depressive disorder will benefit from drugs and/or psychotherapy.

There are two major somatic approaches to the management of depressive episodes: drug treatment and electrotherapy.

Tricyclic antidepressants such as imipramine and amitriptyline are among the most commonly used drugs. They should be administered for at least three or four weeks before shifting to other drugs or to electrotherapy if there is no improvement. The side effects of tricyclic antidepressants are usually not serious. The most common are dry mouth, orthostatic hypotension, and tremor. These effects often diminish as the drug is continued. Less common, but potentially serious, are cardiac arrhythmias from tricyclics.

Individuals vary widely in their ability to metabolize tricyclic antidepressants; as much as a fortyfold difference has been reported between fast and slow metabolizers. What a physician assumes to be a therapeutic dose may actually produce toxic plasma levels or, alternatively, subtherapeutic levels of the drug. For this reason, monitoring plasma levels of tricyclics with gas liquid chromatography (available in many centers and commercial laboratories) is advised when the seemingly therapeutic dose produces either no improvement or potentially serious side effects. There is some evidence that, apart from side effects, high levels of some tricyclics are no more effective than low levels (the "therapeutic window" effect). Monitoring plasma levels may help locate the "therapeutic window."

Lithium also relieves and possibly prevents depression, especially in bipolar depressed patients. It is less useful in unipolar depressed patients (46, 89). About 70 percent of bipolar depressions improve on lithium compared with 40 percent of unipolar depressions (61, 90). If the unipolar patient has a positive family history of bipolar disorder or hypomanic symptoms, lithium may exert a more antidepressant effect than if neither of these features exists (32, 63, 73).

The role of antidepressants in bipolar disorder is not clear. Antidepressants occasionally produce mania in depressed bipolar patients and may indue "rapid cycling" between states of mania and depression. Hence antidepressants, the treatment of choice for unipolar depression, may be deleterious in bipolar depression. A combination of lithium and antidepressants has been suggested for depressed bipolar patients to prevent a sudden onset of mania (61).

Some patients experience depression during certain times of the year, especially in the fall and early winter. DSM-III-R includes Seasonal Affective Disorder (SAD) as a subcategory of Major Depressive Syndrome. Criteria for SAD include a temporal relationship between the onset of bipolar disorder, or recurrent major depression, and a specific period of the year (e.g., regular appearance of depression in the fall).

Seasonal Affective Disorder has been little studied, but a growing body of literature indicates that an effective treatment may exist for the condition. This involves prolonged exposure in the fall and winter months to fluorescent lighting with wavelengths similar to sunlight (96). Devices are now on the market that provide such lighting. There is disagreement about the length of exposure or the time of day when exposure is most efficacious. The effectiveness of the treatment—if definitely shown to exist—may depend on suppressing nighttime melatonin production or some other manipulation of biological rhythms (67). Investigators recommend that patients sit for as long as five hours within a few feet of the light, staring briefly at the light every minute. The possibility of cataracts or retinal damage has not been fully assessed but seems unlikely.

The drug literature is replete with warnings against combining alcohol with psychotropic drugs. In the case of alcohol and barbiturates, the combination may be lethal. There is less additive effect between alcohol and benzodiazepine tranquilizers. Combining alcohol with tricyclic antidepressants may be dangerous. Tricyclics increase the diffusion of alcohol across the blood-brain barrier, resulting in higher levels of alcohol in the brain than in the peripheral circulation (85).

Choosing the most suitable antidepressant for a particular patient is still more of an art than a science. Often the decision is based on side effects rather than evidence of superior effectiveness. For example, amitriptyline causes sedation and is even used as a hypnotic. Imipramine has a more energizing effect and may interfere with sleep. Amitriptyline has marked anticholinergic side effects; desipramine has almost none. When dry mouth or urinary retention are bothersome, desipramine can be substituted for amitriptyline.

Various reports have indicated that urinary excretion of MHPG, a metabolite of norepinephrine, may predict response to specific tricyclics (53). This led to the hypothesis that depression can be biochemically divided into two types. One type is characterized by low urinary MHPG and a favorable response to imipramine or desipramine. The other type involves higher levels of urinary MHPG and a favorable response to amitriptyline. Patients with low levels of urinary MHPG may have decreased central norepinephrine activity. Patients with normal or high urinary levels of MHPG are said to have decreased serotonin activity, which is reflected in lower levels of cerebrospinal fluid (CSF) 5-hydroindoleacetic acid, the major serotonin metabolite. With sufficient confirmation, these reports may yet provide a scientific basis for selecting the most effective antidepressant for a given individual.

Challenge tests with d-amphetamine and methylphenidate may also be useful in selecting the most effective tricyclic. Both of these stimulant drugs, in patients who experience a marked mood elevation from them, predict improvement from imipramine and desipramine, but not from amitriptyline or nortriptyline in several studies (100).

The monoamine oxidase inhibitors are used less frequently than tricyclic antidepressants. There is no evidence that they are more effective, and the side effects can be more serious. Patients taking these drugs should not receive other drugs containing amphetamines or sympathomimetic substances. Foods and beverages containing tyramine (a pressor substance)—particularly cheese, some wines, and beer—should be avoided because of the danger of a hypertensive crisis.

It is not clear whether antidepressant drugs—shown in controlled studies to shorten depressive episodes, reduce the intensity of symptoms; and possibly prevent recurrence—have led to a reduction in suicide by patients with primary affective disorder. The overall suicide rate has not dropped, but there is some evidence that suicide among patients with primary affective disorder has declined, whereas suicide among schizophrenics, alcoholics, homosexuals, and other psychiatric groups has increased (70, 99). Suicide-prevention centers, established in many cities during the 1960s, apparently have had no effect on the suicide rate.

Most psychiatrists reserve electrotherapy for patients who do not respond to antidepressants or who are so ill that they cannot be treated outside a hospital. Electrotherapy may be the most effective form of treatment available for depression (16, 30). When first introduced into clinical practice in the early 1940s, it frequently caused vertebral and other fractures. As advances in the drug modification of electrotherapy have been made, the procedure is less frightening and less commonly associated with complications. Patients are anesthetized briefly with a very rapid-acting barbiturate and then are given a muscle relaxant, usually succinylcholine. Electrodes are placed in the frontotemporal regions and a small, measured amount of electricity is passed between them.

The most troublesome side effect of electrotherapy is the temporary loss of recent memory that occurs in most patients. Delivering unilateral ECT to the nondominant side of the brain minimizes even this sde effect (1). Electrotherapy is probably no more dangerous than treatment with drugs (56). The only absolute contraindication to this mode of therapy is increased intracranial pressure. If required, electrotherapy can be given during pregnancy or postoperatively.

Mania can be treated with phenothiazines, lithium, or electrotherapy. Of these methods, electrotherapy is probably the least effective.

Lithium carbonate may be the drug of choice in the treatment of mania (87), but chlorpromazine is perhaps most useful with very active manic patients. Some clinicians began to treat manics with both chlorpromazine and lithium, stopping chlorpromazine after four or five days when lithium has begun to take effect. If lithium is not effective within ten days, it probably will not be effective at all. Serum levels of lithium required for effective action are in the range of 1 meq/l (88). Generally, total doses of 1200 to 2400 mg of lithium per day in divided form (300 to 600 mg per dose) are required to achieve such serum levels. Because lithium is a potentially toxic drug, its use must be monitored carefully by repeatedly checking serum levels, particularly early in treatment. Some patients experience a fine tremor of the hands at therapeutic levels of lithium (0.8 to 1.5 meq/l). At higher levels, ataxia, disorientation, somnolence, seizures, and finally circulatory collapse may occur. Long-term lithium therapy has also been associated with disturbances of thyroid and renal function, and tubular changes in the kidney have been found on autopsy. Thus, the drug should be used with caution.

The usefulness of lithium may not be limited to the treatment of acute mania. There is some evidence that the drug may reduce morbidity among bipolar patients, preventing depression as well as mania (8), and that it may also prevent depression among some unipolar patients (6, 8, 72).

Two large multicentered trials, evaluating the triazolo benzodiazepine alprazolam in outpatients suffering from major depression, produced encouraging results (37, 92). The antidepressant effect of alprazolam was equal to those of imipramine, amitriptyline, and doxepin. It is not known whether other benzodiazepine-type drugs are also antidepressant, but in one comparison of alprazolam and diazepam, the former was clearly a superior antidepressant (93).

References

1. Abrams, R., Fink, M., Dornbush, R. L., Feldstein, S., Volavka, J., and Roubicek, J. Unilateral and bilateral electroconvulsive therapy. Arch. Gen. Psychiat. 27:88–91, 1972.
2. Abrams, R., and Taylor, M. A. Mania and schizo-affective disorder, manic type: a comparison. Am. J. Psychiat. 133:12, 1976.
3. Abrams, R., and Taylor, M. A. A comparison of unipolar and bipolar depressive illness. Am. J. Psychiat. 137:9, 1084–1087, 1980.
4. Allen, M. G. Twin studies of affective illness. Arch. Gen. Psychiat. 33: 1476–1478, 1976.
5. Amsterdam, J. D., Winokur, A., Caroff, S. N., and Conn, J. The dexamethasone suppression test in outpatients with primary affective disorder and healthy control subjects. Am. J. Psychiat. 139:3, 287–291, 1982.
6. Ananth, J. Treatment approaches to mania. Int. Pharmacopsychiat. 11: 215–231, 1976.
7. Andreasen, N. C., Rice, J., Endicott, J., Coryell, W., Grove, W. M., and Reich, T. Familial rates of affective disorder. Arch. Gen. Psychiat. 44:461–469 1987.
8. Angst, J., Weis, P., Grof, P., Baastrup, P. C., and Schou, J. Lithium prophylaxis in recurrent affective disorders. Brit. J. Psychiat. 116:604–614, 1970.
9. Angst, J. Clinical typology of bipolar illness. In *Mania: An Evolving Concept*, Belmarker, R. H., van Prang, H. M. (eds.), 61–79. Jamaica, N.Y.: SP Medical & Scientific Books, 1980.
10. Baker, M., Dorzab, J., Winokur, G., and Cadoret, R. Depressive disease: the effect of the postpartum state. Biol. Psychiat. 3:357–365, 1971.
11. Beck, A. T. *Depression: Clinical Experimental, and Theoretical Aspects*, New York: Harper & Row, 1967.
12. Beck, A. *Cognitive Therapy and the Emotional Disorders*. New York: International Universities Press, 1976.
13. Becker, J. Achievement related to characteristics of manic-depressives. J. Abnorm. Soc. Psychol. 60:334–339, 1960.
14. Beigel, A., and Murphy, D. L. Unipolar and bipolar affective illness. Arch. Gen. Psychiat. 24:215–220, 1971.
15. Bellack, A. S., Hersen, M., and Himmelhoch, J. Social skills training compared with pharmacotherapy and psychotherapy in the treatment of unipolar depression. Am. J. Psychiat. 138:1562–1567, 1981.
16. Black, D. W., Winokur, G., and Nasrallah, A. The treatment of depression: electroconvulsive therapy v antidepressants: a naturalistic evaluation of 1,495 patients. Compr. Psychiat. 28:169–182, 1987.
17. Blackburn, I. M., Bishop, S., Glen, A. I. M., Whalley, L. J., and Christie, J. E. The efficacy of cognitive therapy in depression: a treatment trial using cognitive therapy and pharmacotherapy, each alone and in combination. Brit. J. Psychiat. 139:181–189, 1981.
18. Bratfos, O., and Haug, J. L. The course of manic-depressive psychosis.

A follow-up investigation of 215 patients. Acta Psychiat. Scand. 44:89–112, 1968.

19. Brockington, I. F., Cernik, K. F., Schofield, E. M., Downing, A. R., Francis, A. F., and Keelan, C. Puerperal psychosis. Arch. Gen. Psychiat. 38:829–833, 1981.

20. Brockington, I. F., Altman, E., Hillier, V. The clinical picture of bipolar affective disorder in its depressed phase. Brit. J. Psychiat. 141:558–562, 1982.

21. Brodie, H. K., and Leff, M. J. Bipolar depression—a comparative study of patient characteristics. Am. J. Psychiat. 127:1086–1090, 1971.

22. Carroll, B. J. The dexamethasone test for melancholia. Brit. J. Psychiat. 140:292–304, 1982.

23. Cassidy, W. L., Flanigan, M. B., Spellman, M., and Cohen, M. E. Clinical observations in manic depressive disease. A quantitative study of 100 manic depressive patients in 50 medically sick controls. JAMA 164: 1535–1546, 1953.

24. Clayton, P. J., Rodin, L., and Winokur, G. Family history studies, III. Schizo-affective disorder, clinical and genetic factors including a one to two year follow-up. Compr. Psychiat. 9:31–49, 1968.

25. Clayton, P. J., Halikas, J. A., and Maurice, W. L. The bereavement of the widowed. Dis. Nerv. Syst. 32:597–604, 1971.

26. Clayton, P. J., Halikas, J. A., and Maurice, W. L. The depression of widowhood. Brit. J. Psychiat. 120:71–78, 1972.

27. Clayton, P. J. Schizoaffective disorders. J. Nerv. Ment. Dis. 170:646–650, 1982.

28. Cohen, M. B., Baker, G., Cohen, R. A., Fromm-Reichmann, F., and Weigert, E. V. An intensive study of 12 cases of manic depressive psychosis. Psychiatry 17:103–137, 1954.

29. Cohen, S. M., Allen, M. G., Pollin, W., and Hrubec, Z. Relationship of schizo-affective psychosis to manic depressive psychosis and schizophrenia. Arch. Gen. Psychiat. 26:539–546, 1972.

30. Davis, J. M. Efficacy of tranquilizing and antidepressant drugs. Arch. Gen. Psychiat. 13:552–572, 1965.

31. DiMascio, A., Weissman, M. M., Prusoff, B. A., Neu, C., Zwilling, M., and Klerman, G. L. Differential symptom reduction by drugs and psychotherapy in acute depression. Arch. Gen. Psychiat. 36:1450–1456, 1979.

32. Dunner, D. L., Stallone, F., and Fieve, R. R. Lithium carbonate and affective disorders. Arch. Gen. Psychiat. 33:117–120, 1976.

33. Dunner, D. L., and Fieve, R. R. The effect of lithium in depressive subtypes. In Neuropsychopharmacology, Deniker, A., Radouco-Thomas, B., and Villeneuve, C., (eds.), 1109–1115. New York: Pergamon Press, 1978.

34. Dunner, D. L. Unipolar and bipolar depression: recent findings from clinical and biologic studies. In Psychobiology of Affective Disorders, 11–24. New York: Plenum, 1980.

35. Egeland, J. A., and Hostetter, A. M.   Amish study, I: affective disorders among the Amish, 1976–1980. Am. J. Psychiat. 140:1, 56–61, 1983.
36. Egeland, J. A., Gerhard, D. S., Pauls, D. L., Sussex, J. N., Kidd, K. K., Allen, C. R., Hostetter, A. M., and Housman, D. E.   Bipolar affective disorders linked to DNA markers on chromosome 11. Nature 325:783, 1987.
37. Feighner, J. P., Aden, G. C., Fabre, L. F., Rickels, K., and Smith, W. T. Comparison of alprazolam, imipramine, and placebo in the treatment of depression. JAMA 249:3057–3064 1983.
38. Ferster, C. B.   Behavioral approaches to depression. In *The Psychology of Depression: Contemporary Theory and Research*, Friedman, R. J., and Katz, M. M. (eds.), 29–54. New York: John Wiley & Sons, 1974.
39. Fremming, K.   The expectation of mental infirmity in the sample of the Danish population. In *Occasional Papers of Eugenics*, no. 7. London: Cassell, 1951.
40. Freud, S.   *Mourning and Melancholia*. In *The Complete Psychological Works of Sigmund Freud*, 14:243–258. London: Hogarth Press, 1957.
41. Gibson, R. W.   The family background and early life experience of the manic depressive patient. Psychiatry 21:71–90, 1968.
42. Gold, M. S., Pottash, A. L. C., Extein, I., and Sweeney, D. R.   Diagnosis of depression in the 1980's. JAMA 245:1562–1564, 1981.
43. Goldin, L. R., Clerget-Darpoux, F., and Gershon, E. S.   Relationship of HLA to major affective disorder not supported. Psychiat. Res. 7:29–45, Elsevier Biomedical Press, 1982.
44. Goodwin, F. K., Murphy, D. L., and Dunner, D. L.   Lithium response in unipolar versus bipolar depression. Am. J. Psychiat. 129:44–47, 1974.
45. Goodwin, D. W., Schulsinger, F., Knop, J., Mednick, S., and Guze, S. B. Psychopathology in adopted and nonadopted daughters of alcoholics. Arch. Gen. Psychiat. 34:751–755, 1977.
46. Goodwin, F. K., Murphy, D. L., Dunner, D. L., and Bunney, W. E. Lithium response in unipolar vs. bipolar depression. Am. J. Psychiat. 129:44–47, 1972.
47. Greden, J. F., Albala, A. A., Haskett, R. F., et al.   Normalization of dexamethasone suppression test: a laboratory index of recovery from endogenous depression. Biol. Psychiat. 15:449–458, 1980.
48. Guze, S. B., and Robins, E.   Suicide and primary affective disorders. Brit. J. Psychiat. 117:437–438, 1970.
49. Guze, S. B., Woodruff, R. A., and Clayton, P. J.   Secondary affective disorder: a study of 95 cases. Psychol. Med. 1:426–428, 1971.
50. Helgason, T.   Epidemiology of mental disorders in Iceland. A psychiatric and demographic investigation of 5395 Icelanders. Acta Psychiat. Scand. 40, Supply. 173, 1964.
51. Hirschfeld, R. M. A., and Cross, C. K.   Epidemiology of affective disorders. Arch. Gen. Psychiat. 39:35–46, 1982.
52. Hodgkinson, S., Sherrington, R., Gurling, H., Marchbanks, R., Reed-

ers, S., Mallet, J., McInnis, M., Petursson, H., and Brynjolfsson, J.
Molecular genetic evidence for heterogeneity in manic depression. Nature 325:805, 1987.
53. Hollister, L. E., Davis, K. L., and Berger, P. A. Subtypes of depression
based on excretion of MHPG and response to nortriptyline. Arch. Gen.
Psychiat. 37:1107–1110, 1980.
54. James, W. The Varieties of Religious Experience. In The Epidemiology
of Depression, Silverman, C. (ed.). Baltimore: Johns Hopkins Univ.
Press, 1968.
55. Johnson, G. F. S., and Leeman, M. M. Onset of illness in bipolar
manic-depressives and their affectively ill first-degree relatives. Biol. Psychiat. 12:733–741, 1977.
56. Kalinowsky, L. B., and Hippius, H. Pharmacological, Convulsive and
Other Somatic Treatments in Psychiatry. New York: Grune & Stratton,
1969.
57. Keller, M. B., Shapiro, R. W., Lavori, P. W., and Wolfe, N. Recovery
in major depressive disorder. Arch. Gen. Psychiat. 39:905–910, 1982.
58. Keller, M. B., Shapiro, R. W., Lavori, P. W., and Wolfe, N. Relapse
in major depressive disorder. Arch. Gen. Psychiat. 39:911–915, 1982.
59. Kendell, R. E. The Classification of Depressive Illnesses, Maudsley
Monogr. no. 18. London: Oxford Univ. Press, 1968.
60. Kerr, T. A., Schapira, K., and Roth, M. The relationship between premature death and affective disorders. Brit. J. Psychiat. 115:1277–1282,
1969.
61. Klein, D. F., Gittelman, R., and Quitkin, F. Review of the literature
on mood-stabilizing drugs. In Diagnosis and Drug Treatment of Psychiatric Disorders: Adults and Children, 268–408. Baltimore: Williams
& Wilkins, 1980.
62. Kraepelin, E. Manic Depressive Insanity and Paranoia. Edinburgh: E. S.
Livingstone, 1921.
63. Kupfer, D. J., Pikar, D., and Himmelhoch, J. M. Are there two types
of unipolar depression? Arch. Gen. Psychiat. 32:866–871, 1975.
64. Kupfer, D. J., and Foster, F. G. EEG sleep and depression. In Sleep
Disorders: Diagnosis and Treatment, Williams, R. L., and Karacan, I.
(eds.), 163–204. New York: John Wiley & Sons, 1978.
65. Leonhard, K., Korff, I., and Shulz, H. Die Temperamente in den Fa-
milien der monopolaren und bipolaren phasichen Psychosen. Psychiat.
Neurol. 143:416–434, 1962.
66. Lewis, A. Melancholia: a historical review. In The State of Psychiatry:
Essays and Addresses. New York: Science House, 1967.
67. Lewy, A., Sack, R., and Singer, C. Antidepressant and circadian phase
shifting effects of light. Science 235:352–354, 1987.
68. Lundquist, G. Prognosis and course in manic depressive psychosis. A
follow-up study of 319 first admissions. Acta Psychiat. et Neurol., Suppl.
35, 1945.
69. Marcotte, D. B., Kane, F. G., Obrist, P., and Lipton, M. A. Psycho-

physiologic changes accompanying oral contraceptive use. Brit. J. Psychiat. 116:165–167, 1970.

70. Martin, R. L., Cloninger, C. R., Guze, S. B., and Clayton, P. J. Mortality in 500 psychiatric outpatients. Arch. Gen. Psychiat., 1983.
71. Matthysse, S., and Kidd, K. K. Evidence of HLA linkage in depressive disorders. N. Engl. J. Med. 305:1340–1341, 1981.
72. Mendels, J., Secunda, S. K., and Dyson, W. L. The controlled study of the antidepressant effects of lithium carbonate. Arch. Gen. Psychiat. 26: 154–157, 1972.
73. Mendels, J., Ramsey, A., and Dyson, W. L. Lithium as an antidepressant. Arch. Gen. Psychiat. 36:845–846, 1979.
74. Mendlewicz, J., and Rainer, J. D. Adoption study supporting genetic transmission in manic-depressive illness. Nature 268:327–329, 1977.
75. Mendelwicz, J., Simon, P., Sevy, S., Charon, F., Brocas, H., Legros, S., and Vassart, G. Polymorphic DNA marker on X chromosome and manic depression. Lancet, 1230, May 30, 1987.
76. Mood Disorders: Pharmacologic Prevention of Recurrences. Public Health Service Monogr., vol. 5, no. 4, 1986.
77. Murphy, G. E., and Wetzel, R. D. Family history of suicidal behavior among suicide attempters. J. Nerv. Ment. Dis. 170:2, 86–90, 1982.
78. Myers, J. K., Weissman, M. M., and Tischler, G. C. Six-month prevalence of psychiatric disorders in three communities. Arch. Gen. Psychiat. 41:959–967, 1984.
79. Nicholi, A. M. Harvard dropouts: some psychiatric findings. Am. J. Psychiat. 124:651–658, 1967.
80. Perris, C. A study of bipolar (manic-depressive) and unipolar recurrent depressive psychoses. Acta Psychiat. Scand., Suppl. 194, 1966.
81. Perris, C. The distinction between bipolar and unipolar affective disorders. In Handbook of Affective Disorders, Paykel, E. S. (ed.), 45–48. New York: Guilford Press, 1982.
82. Peselow, E. D., Goldring, N., Fieve, R. R., and Wright, R. The dexamethasone suppression test in depressed outpatients and normal control subjects. Am. J. Psychiat. 140:2, 1983.
83. Pierce, D. W. The predictive validation of a suicide intent scale: a five year follow-up. Brit. J. Psychiat. 139:391–396, 1981.
84. Poznaski, E., and Zrull, J. P. Childhood depression. Clinical characteristics of overtly depressed children. Arch. Gen. Psychiat. 23:8–15, 1970.
85. Preskorn, S. H., Irwin, G. H., Simpson, S., Friesen, D., Rinne, J., and Jerkovich, G. Medical therapies for mood disorders alter the blood–barrier. Science 213:469–471, 1981.
86. Price, J. S. Chronic depressive illness. Brit. Med. J. 1:1200–1201, 1978.
87. Prien, R. F., Caffey, E. M., and Klett, C. J. Comparison of lithium carbonate and chlorpromazine in the treatment of mania. Arch. Gen. Psychiat. 26:146–153, 1972.
88. Prien, R. F., Caffey, E. M., and Klett, C. J. Relationship between

serum lithium level and clinical response in acute mania treated with lithium. Brit. J. Psychiat. 120:409–414, 1972.

89. Prien, R. F., Kupfer, D. J., and Mansky, P. A. Drug therapy in the prevention of recurrences in unipolar and bipolar affective disorders. Arch. Gen. Psychiat. 41:1096–1104, 1984.

90. Ramsey, T. A., and Mendels, J. Lithium in the acute treatment of depression. In *Handbook of Lithium Therapy*, Johnson, F. N. (ed.), 17–25. Lancaster, England, MTP Press, 1980.

91. Reich, T., and Winokur, G. Postpartum psychoses in patients with manic depressive disease. J. Nerv. Ment. Dis. 151:60–68, 1970.

92. Rickels, K., Feighner, J. P., and Smith, W. T. Alprazolam, amitriptyline, doxepin, and placebo in the treatment of depression. Arch. Gen. Psychiat. 42:134–141, 1985.

93. Rickels, K., Chung, H. R., Csanalosi, I. B., Hurowitz, A. M., London, J., Wiseman, K., Kaplan, M., and Amsterdam, J. D. Alprazolam, diazepam, imipramine, and placebo in outpatients with major depression. Arch. Gen. Psychiat. 44:862–866, 1987.

94. Robins, E., and Guze, S. B. Classification of affective disorders: the primary-secondary, the endogeneous-reactive, and the neurotic-psychotic concepts. In *Recent Advances in Psychobiology of the Depressive Illnesses*, Proceedings of a workshop sponsored by the NIMH, Williams, T. A., Katz, M. M., and Shield, J. A. (eds.). U.S. GPO, 1972.

95. Robins, E., Murphy, G. E., Wilkinson, R. H., Gassner, S., and Kayes, J. Some clinical considerations in the prevention of suicide based on a study of 134 successful suicides. Am. J. Public Health 49:888–899, 1959.

96. Rosenthal, N., Sack, D., and Gillin, J. Seasonal affective disorder: a description of the syndrome and preliminary findings with light therapy. Arch. Gen. Psychiat. 41:72–80, 1985.

97. Roth, M. Natural history of mental disorder in old age. J. Ment. Sci. 101:281–301, 1955.

98. Rounsaville, B. J., Klerman, G. L., and Weissman, M. M. Do psychotherapy and pharmacotherapy for depression conflict? Arch. Gen. Psychiat. 38:24–29, 1981.

99. Roy, A. Risk factors for suicide in psychiatric patients. Arch. Gen. Psychiat. 39:1089–1095, 1982.

100. Sabelli, H. C., Fawcett, J., Javaid, J. I., and Bagri, S. The methylphenidate test for differentiating desipramine-responsive from nortriptyline-responsive depression. Am. J. Psychiat. 140:2, 212–217, 1983.

101. Saccetti, E., Vita, A., Conte, G. Lithium prophylaxis in major affective disorders: on the specificity and sensitivity of some "new" predictors of treatment outcome. In *Current Trends in Lithium and Rubidium Therapy*, Corsini, G. V. (ed.), 165–187. Lancaster, England, MTP Press, 1984.

102. Savard, R. J., Rey, A. C., and Post, R. M. Halstead-Reitan Category Test in bipolar and unipolar affective disorders. J. Nerv. Ment. Dis. 168:297–304, 1980.

103. Schildkraut, J. J., Orsulak, P. J., and Schatzberg, A. F. Toward a biochemical classification of depressive disorders: I. Differences in urinary excretion of MHPG and other catecholamine metabolites in clinically defined subtypes of depressions. Arch. Gen. Psychiat. 35:1427–1433, 1978.

104. Schildkraut, J. J., Orsulak, P. J., and LaBrie, R. A. Toward a biochemical classification of depressive disorders: II. Application of multivariate discriminant function analysis to data on urinary catecholamines and metabolites. Arch. Gen. Psychiat. 35:1436–1439, 1978.

105. Schildkraut, J. J., Schatzberg, A. F., and Orsulak, P. J. Biological discrimination of subtypes of depression. In The Affective Disorders, Davis, J. M., and Maas, J. W. (eds.), 31–51. Washington, D.C.: American Psychiatric Press, 1983.

106. Schwartz, J. M., Baxter, L. R., Mazziotta, J. C., Gerner, R. H., and Phelps, M. E. The differential diagnosis of depression. Relevance of positron emission tomography studies of cerebral glucose metabolism to the bipolar-unipolar dichotomy. JAMA 258:1368–1374, 1987.

107. Shobe, F. O., and Brione, P. Long-term prognosis in manic-depressive illness. A follow-up investigation of 111 patients. Arch. Gen. Psychiat. 24:334–337, 1971.

108. Silverman, C. (ed.). The Epidemiology of Depression. Baltimore: Johns Hopkins Univ. Press, 1968.

109. Sternberg, D. E., and Jarvik, M. E. Memory functions in depression. Arch. Gen. Psychiat. 33:219–224, 1976.

110. Strömgren, E. Contributions to psychiatric epidemiology and genetics. Acta Jutlandica Med. Series 16, vol. 40, 1968.

111. Suarez, B. K., and Croughan, J. Is the major histocompatibility complex linked to genes that increase susceptibility to affective disorder? A critical appraisal. Psychiat. Res. 7:19–27, Elsevier Biomedical Press, 1982.

112. Taylor, F. G., and Marshall, W. L. Experimental analysis of a cognitive-behavioral therapy for depression. Cognitive Therapy Res. 1:59–72, 1977.

113. APA Task Force on Laboratory Tests in Psychiatry. The dexamethasone suppression test: an overview of its current status in psychiatry. Am. J. Psychiat. 144:10, 1987.

114. Vaillant, G. E. Manic-depressive heredity and remission in schizophrenia. Brit. J. Psychiat. 109:746–749, 1963.

115. VanValkenburg, C., Lowry, M., Winokur, G., and Cadoret, R. Depression spectrum disease versus pure depressive disease. J. Nerv. Ment. Dis. 165:341–347, 1977.

116. Weissman, M. M., Klerman, G. L., Paykel, E. S., Prusoff, B., and Hanson, B. Treatment effects on the social adjustment of depressed patients. Arch. Gen. Psychiat. 30:771–778, 1974.

117. Weissman, M. M., and Klerman, G. L. Sex differences and the epidemiology of depression. Arch. Gen. Psychiat. 34:98–111, 1977.

118. Weissman, M. M., and Myers, J. K. Affective disorders in a US urban community. Arch. Gen. Psychiat. 35:1304–1311, 1978.

119. Weissman, M. M., Myers, J. K., and Harding, P. S.   Psychiatric disorders in a U.S. urban community: 1975–76. Am. J. Psychiat. 135:459–462, 1978.

120. Weissman, M. M.   The psychological treatment of depression. Arch. Gen. Psychiat. 36:1261–1269, 1979.

121. Weitkamp, L. R., Stancer, H. C., Persad, E., Flood, C., and Guttormsen, S.   Depressive disorders and HLA: A gene on chromosome 6 that can affect behavior. N. Engl. J. Med. 305:1301–1306, 1981.

122. Whitney, W., Cadoret, R. J., and McClure, J. N.   Depressive symptoms and academic performance in college students. Am. J. Psychiat. 128:766–770, 1971.

123. Winokur, G., and Pitts, F. N.   Affective disorder: VI. A family history study of prevalences, sex differences, and possible genetic factors. J. Psychiat. Res. 3:113–123, 1965.

124. Winokur, G.   Types of depressive illness. Brit. J. Psychiat. 120:265–266, 1972.

125. Woodruff, R. A., Murphy, G. E., and Herjanic, M.   The natural history of affective disorders—I. Symptoms of 72 patients at the time of index hospital admission. J. Psychiat. Res. 5:255–263, 1967.

126. Woodruff, R. A., Robins, L. N., Winokur, G., and Reich, T.   Manic depressive illness and social achievement. Acta Psychiat. Scand. 47:237–249, 1971.

127. Woodruff, R. A., Guze, S. B., and Clayton, P. J.   Anxiety neurosis among psychiatric outpatients. Compr. Psychiat. 13:165–170, 1972.

128. Woodruff, R. A., Guze, S. B., Clayton, P. J., and Carr, D.   Alcoholism and depression. Arch. Gen. Psychiat. 28:97–100, 1973.

129. Woodruff, R. A., Guze, S. B., and Clayton, P. J.   Alcoholics who see a psychiatrist compared with those who do not. Q. J. Stud. Alcohol 34:1162–1171, 1973.

130. Young, W., Knowles, J. B., MacLean, A. W., Boag, L., and McConville, B. J.   The sleep of childhood depressives: comparison with age-matched controls. Biol. Psychiat. 17:10, 1163–1168, 1982.

2. Schizophrenic Disorders

Definition

Hallucinations and delusions are considered hallmarks of mental disorder and thus are of great interest to psychiatrists. These symptoms may be seen in a wide variety of illnesses, including primary affective disorders, brain syndromes, alcoholism, drug dependence, and a group of conditions that may loosely be called the schizophrenic disorders. Many investigators believe that schizophrenic disorders comprise a number of different conditions, but efforts to divide them into valid subgroups generally have not been successful and inconsistent usage has made nomenclature confusing. Extensive work, however, has indicated that the schizophrenic disorders may be divided into two major categories: one with a relatively poor prognosis, the other with a relatively good prognosis.

When this differentiation is attempted, such labels as schizophrenia, chronic schizophrenia, process schizophrenia, nuclear schizophrenia, and nonremitting schizophrenia are used to refer to the poor prognosis cases, whereas schizophreniform, acute schizophrenia, reactive schizophrenia, schizoaffective, and remitting schizophrenia are used for the good prognosis cases (108).

The schizophrenic disorders are manifested, at least intermittently, by delusions and hallucinations in a clear sensorium. Other noteworthy features are a blunted, shallow, or strikingly inappropriate affect; odd, sometimes bizarre, motor behavior (termed "catatonic"); and disordered thinking in which goal-directedness and normal associations between ideas are markedly distorted (loosening of associations and tangential thinking).

Although the schizophrenic disorders are generally chronic, some cases appear as episodic illnesses and tend to develop in individuals whose previous general adjustment has been good. These patients are likely to show striking affective symptoms during the illness. They may seem perplexed and bewildered and may be mildly disoriented. Though recovery from the psychotic episode is usual, repeated episodes are common. Subsequent episodes may be less pronounced and may sometimes resemble affective disorders (140, 141).

Historical Background

In now classic studies, Emil Kraepelin (1856–1926), a German psychiatrist, building upon the work of his countrymen K. L. Kahlbaum (1828–99) and E. Hecker (1843–1909), who described "catatonia" and "hebephrenia," respectively, laid the groundwork for present views of schizophrenia (74). After careful follow-up of hospitalized patients, he separated "manic depressive psychosis" from "dementia praecox." The latter term referred to the disorder now called "chronic schizophrenia." Even though Kraepelin believed dementia praecox to be a chronic disorder that frequently ended in marked deterioration of the personality, he recognized that a small number of patients recovered completely. His "narrow" view of schizophrenia has been followed by most European psychiatrists, particularly those in the Scandinavian countries and Great Britain.

A "broader" approach to schizophrenia (and the name itself) was offered by Eugen Bleuler (1857–1939), a Swiss psychiatrist, who realized that he might be dealing with a group of disorders (15). His diagnostic criteria were not based on their ability to predict course and outcome, but on their conforming to his hypothesis concerning the basic defect, namely, the "splitting" of psychic functions. By this he meant inconsistency, inappropriateness, and disorganization of affect, thought, and action, in the absence of obvious brain disease. Despite his less strict approach to diagnosis and the variable course that his patients experienced, Bleuler believed that patients with schizophrenia never recovered completely, never returned to their premorbid state (*restitutio ad integrum*). Because his "fundamental" symptoms included autism (defined by Bleuler as "divorce from reality"), blunted or inappropriate affect, ambivalence, and disturbed association of thought—all often difficult to define and specify—Bleuler's work set the stage for the very broad concepts of schizophrenia developed by psychoanalysts and adopted by many American psychiatrists.

Psychoanalytic theory views schizophrenia primarily as a manifestation of a "weak ego." Unable to cope with the problems of life and unable to use effectively the "defenses of the ego" to handle instinctual forces and anxiety, patients "regress" to a primitive psychosexual level of functioning ("primary process") manifested by thought disorder, affective poverty, disorganization, and inability to conform to the demands of "reality" (19). In this psychoanalytic view of schizophrenia, all evidence of weak ego (including a wide range of personality handicaps and abnormalities) or of primary process (such as hallucinations, delusions, poor reality-testing, tangential thinking, and ambivalence) may be a manifestation of schizophrenia. It is not surprising, therefore, that the diagnosis was used in a wide range of clinical situations.

In the late 1930s a number of European and American investigators began again to approach the problem of the schizophrenias in terms of predicting course, response to treatment, and long-term outcome. They were influenced first by the advent of electroconvulsive therapy (ECT) and later by the introduction of the phenothiazines. Langfeldt (77), Astrup (7, 8), Retterstöl (102, 103), Leonhard (80), Ey (38), Stephens (127, 128), and Vaillant (140, 141) have been leaders in these efforts.

Proper evaluation of treatment requires a knowledge of the natural history of the disorder being treated, especially the factors associated with different clinical courses and outcomes. Such factors can only be identified by follow-up studies such as those carried out by the above investigators. Thus, current clinical and research approaches to the schizophrenic disorders are based on extensive follow-up studies that are supplemented by family studies (108).

Epidemiology

Chronic schizophrenia occurs in somewhat less than 1 percent of the population, but because of its early onset, chronicity, and associated disability, it is one of the most important psychiatric illnesses. It has been estimated that between one-third and one-half of all psychiatric beds in U.S. hospitals are occupied by such patients. The distinction between good and poor prognosis forms of schizophrenia often is not made in epidemiologic surveys, nor is the possibility raised that some of the good prognosis cases may be other disorders, though available evidence suggests that the good prognosis cases are more common than the poor prognosis ones (103). The combined prevalence of both good and poor prognosis disorders is probably between 1 and 2 percent.

Schizophrenic disorders are found in all cultures. A number of studies have indicated that they are more prevalent among people from lower socioeconomic backgrounds. For some investigators, this means that poverty, limited education, and associated handicaps predispose to schizophrenic illness. Studies have shown, however, that the association between schizophrenic illness and low socioeconomic status can be explained by "downward drift," a term that refers to the effect of an illness on the patients' socioeconomic status. If a disorder interferes with education and work performance, so that individuals are not able to complete advanced schooling or hold positions of responsibility, their socioeconomic status—characteristically defined by income, educational achievement, and job prestige—cannot be high.

Studies in England (45), Denmark (145), Finland (114), and the United States (31) have shown that the distribution of socioeconomic class among the fathers of schizophrenics is the same as in the general population, indicating that the lower socioeconomic status of patients with schizophrenic illness is the result of downward drift.

Another interesting observation concerning the epidemiology of schizophrenia is the reported tendency of schizophrenics in Europe and the United States to be born during the late winter and early spring months of the year (57, 58, 138). Though most schizophrenics are not born during these months, the data suggest that something associated with such births predisposes to schizophrenic illness. Other work suggests a similar association with bipolar affective disorder (18), raising the likelihood that the seasonal effect may be nonspecific. These reports do not distinguish between good prognosis and poor prognosis cases, so they leave the reader uncertain as to whether the observation is equally applicable to both forms of the disorder.

Clinical Picture

Common delusions in schizophrenia are those of persecution and control in which patients believe others are spying on them, spreading false rumors about them, planning to harm them, trying to control their thoughts or actions, or reading their minds.

A young woman complained bitterly that her brother was sending special mysterious messages to her by means of television in order to make her do things that would call attention to her and lead to trouble with the police. A young man was convinced that he was being followed and observed on the streets and in various buildings, but he accepted this as being directed by his psychiatrist as a way to monitor a patient's progress. Patients may express the belief that they are the victims of conspiracies by Communists, Catholics, neighbors, the FBI, and so on. Delusions of depersonalization are also common. These may be feelings that bizarre bodily changes are taking place, sometimes as a result of the deliberate but obscure actions of others. "My insides are rotting because *they* are poisoning my food. It's because *they* know I'm wise to them and have reported them to the police."

The most common hallucinations are auditory. They may involve solitary or multiple voices. The patient may or may not recognize the voices or talk back to them. The voices may seem to come from within the patient's body or from outside sources such as radios or walls. The voices may criticize, ridicule, or threaten; often they urge the patient to do something she or he believes is wrong. Visual hallucinations are also frequent (47). These may vary from frightening vague forms to images of dead or absent relatives to scenes of violence or hell. Olfactory hallucinations, which are infrequent, usually consist of unpleasant smells arising from the patient's own body. Tactile (haptic) hallucinations, also infrequent, may consist of feelings that one's genitals are being manipulated, that there are animals inside one's body, or that there are insects crawling over one's skin.

Though the so-called "typical" flat schizophrenic affect is highly characteristic when it is severe, its diagnostic value is limited because it frequently is subtle, leading to disagreement about its presence. Even when clearly present, it is not easy to describe. Patients seem emotionally unresponsive, without warmth or empathy. They can talk about frightening or shocking thoughts without seeming to experience their usual emotional impact ("inappropriate" affect). It is often difficult to feel compassion and sympathy for the patient or to believe that he can empathize with others.

Recurrent posturing, grimacing, prolonged immobility, and "waxy flexibility" are dramatic examples of catatonic behavior. These may be independent symptoms or responses to auditory hallucinations.

The impaired goal-directedness of schizophrenic thought and speech may take various forms, all likely to occur in the same patient: blocking, in which the patient's thought and speech stop for periods of time only to begin again with an apparently different subject; tangential associations, in which connections between thoughts are difficult or impossible to follow; neologisms, in which the patient makes up new words; or "word salad," in which the patient's speech consists of words without any understandable sequence or meaning.

Following Kraepelin and Bleuler (15, 74), many psychiatrists group schizophrenics into paranoid, catatonic, hebephrenic, and simple types, depending on whether the predominant symptoms are delusions, bizarre motor behavior, disturbances in affect and association, or social withdrawal and inadequacy. In practice, however, symptoms vary with time, so that a patient may seem to fit several of the subclassifications during the course of his or her illness (20, 70). Attempts to identify delusions or hallucinations characteristic of good or poor prognosis cases have not been consistently successful (47).

Patients with schizophrenic disorders may display prominent alterations of mood, usually depression but sometimes euphoria, during the course of illness (100). Other affective symptoms—such as insomnia, anorexia, weight loss, alterations in interest and energy, impairment of mental concentration, guilt, and suicidal preoccupation—may also be present. Often, particularly early in the course of good prognosis cases, the patients may appear confused, perplexed, somewhat disorientated. In fact, a substantial proportion of schizophrenic patients experience episodes of depression that symptomatically resemble those seen in primary affective disorder (53). Because the first-degree relatives of such patients do not have an increased prevalence of affective disorders, their depression may be regarded as symptomatic rather than the manifestation of a second illness.

The introduction of a variety of brain-imaging techniques, including computerized tomography (CT), magnetic resonance, and positron emission tomography (PET), has led to a renewed interest in the possibility of identifying structural abnormalities in the brains of schizophrenic patients. A number of publications suggest that some patients have an increased cerebral-ventricle to brain-tissue ratio and that this may be correlated in these patients with evidence of cognitive impairment on standard neuropsychological tests. The reported percentage of such abnormalities has varied greatly in different studies and some authors have been unable to find any significant differences between schizophrenics and controls, but the findings seem important and further investigations are under way (5, 6, 29, 82, 91, 92, 104, 105, 133). A perhaps related observation from recent postmortem and CT studies is that some schizophrenic patients show a smaller cerebellar vermis than control subjects (81, 144). Finally, a number of reports indicate that schizophrenic patients may have specific abnormalities in structure and function of other brain regions, including the cerebral hemispheres, the corpus callosum, and the basal ganglia (3, 4, 12, 13, 16, 22, 51, 52, 87, 90). An especially interesting finding suggests that in young, never-medicated schizophrenics there is an abnormally high blood flow in the left globus pallidus (32). Another finding of elevated D_2 dopamine receptors in drug-naive subjects may be related to this (146). Although the validity and significance of these findings must await further research, it is clear that many schizophrenic patients have unusual or abnormal brain structure or function.

It has also been reported that some schizophrenic patients have lowered platelet monoamine oxidase (MAO) activity (149). This is of interest because of the many apparent similarities between platelets and brain synaptosomes, but its significance is uncertain because similar differences in brain MAO activity have not been observed at postmortem examination (106). Of perhaps greater interest are recent studies based on postmortem examination of brains of schizophrenic patients that suggest a variety of subtle anatomical changes such as differential neuronal density in different areas and layers of the cerebral cortex (10, 21). Other studies of brain function include an exciting report of lateral asymmetry in amygdala dopamine in schizophrenics (107). This indicates that postmortem tissue can be used for such studies despite neuroleptic drug use as the asymmetry should be independent of drug use. Finally, the possibility of a viral etiology for schizophrenia, based on elevated antibody titers, has again attracted interest (26).

For many years it has been suggested that some schizophrenic patients, particularly those whose illness begins later in life with prominent paranoid delusions, may have a greater degree of hearing loss than controls (33). The role of such hearing loss in the pathogenesis of auditory hallucinations is not yet clear.

Some psychiatrists, at least for research purposes, emphasize the difference between "positive" symptoms (delusions, hallucinations, bizarre behavior) and "negative" symptoms (apathy, withdrawal, unresponsiveness); others have challenged the usefulness of such a distinction.

The criteria for diagnosing schizophrenia according to DSM-III-R are presented in Table 2.1. Other criteria have been pro-

Table 2.1 Diagnostic criteria for Schizophrenia (DSM-III-R)

A. Presence of characteristic psychotic symptoms in the active phase: either (1), (2), or (3) for at least one week (unless the symptoms are successfully treated):
   (1) two of the following:
       (a) delusions
       (b) prominent hallucinations (throughout the day for several days or several times a week for several weeks, each hallucinatory experience not being limited to a few brief moments)
       (c) incoherence or marked loosening of associations
       (d) catatonic behavior
       (e) flat or grossly inappropriate affect
   (2) bizarre delusions (i.e., involving a phenomenon that the person's culture would regard as totally implausible, e.g., thought broadcasting, being controlled by a dead person)
   (3) prominent hallucinations [as defined in (1)(b) above] of a voice with content having no apparent relation to depression or elation, or a voice keeping up a running commentary on the person's behavior or thoughts, or two or more voices conversing with each other
B. During the course of the disturbance, functioning in such areas as work, social relations, and self-care is markedly below the highest level achieved before onset of the disturbance (or, when the onset is in childhood or adolescence, failure to achieve expected level of social development).
C. Schizoaffective Disorder and Mood Disorder with Psychotic Features have been ruled out, i.e., if a Major Depressive or Manic Syndrome has ever been present during an active phase of the disturbance, the total duration of all episodes of a mood syndrome has been brief relative to the total duration of the active and residual phases of the disturbance.
D. Continuous signs of the disturbance for at least six months. The six-month period must include an active phase (of at least one week, or less if symptoms have been successfully treated) during

which there were psychotic symptoms characteristic of Schizophrenia (symptoms in A), with or without a prodromal or residual phase, as defined below.

Prodromal phase: A clear deterioration in functioning before the active phase of the disturbance that is not due to a disturbance in mood or to a Psychoactive Substance Use Disorder and that involves at least two of the symptoms listed below.

Residual phase: Following the active phase of the disturbance, persistence of at least two of the symptoms noted below, these not being due to a disturbance in mood or to a Psychoactive Substance Use Disorder.

Prodromal or Residual Symptoms:

(1)   marked social isolation or withdrawal
(2)   marked impairment in role functioning as wage-earner, student, or homemaker
(3)   markedly peculiar behavior (e.g., collecting garbage, talking to self in public, hoarding food)
(4)   marked impairment in personal hygiene and grooming
(5)   blunted or inappropriate affect
(6)   digressive, vague, overelaborate, or circumstantial speech, or poverty of speech, or poverty of content of speech
(7)   odd beliefs or magical thinking, influencing behavior and inconsistent with cultural norms, e.g., superstitiousness, belief in clairvoyance, telepathy, "sixth sense," "others can feel my feelings," overvalued ideas, ideas of reference
(8)   unusual perceptual experiences, e.g., recurrent illusions, sensing the presence of a force or person not actually present
(9)   marked lack of initiative, interests, or energy

Examples: Six months of prodromal symptoms with one week of symptoms from A; no prodromal symptoms with six months of symptoms from A; no prodromal symptoms with one week of symptoms from A and six months of residual symptoms.

E.   It cannot be established that an organic factor intiated and maintained the disturbance.

F.   If there is a history of Autistic Disorder, the additional diagnosis of Schizophrenia is made only if prominent delusions or hallucinations are also present.

Classification of course. The course of the disturbance is coded in the fifth digit:

1-Subchronic. The time from the beginning of the disturbance, when the person first began to show signs of the disturbance (including prodromal, active, and residual phases) more or less continuously, is less than two years, but at least six months.

2-Chronic. Same as above, but more than two years.

3-Subchronic with Acute Exacerbation. Reemergence of prominent psychotic symptoms in a person with a subchronic course course who has been in the residual phase of the disturbance.

4-Chronic with Acute Exacerbation. Reemergence of prominent psychotic symptoms in a person with a chronic course who has been in the residual phase of the disturbance.

5-In Remission. When a person with a history of Schizophrenia is free of all signs of the disturbance (whether or not on medication), "in Remission" should be coded. Differentiating Schizophrenia in Remission from No Mental Disorder requires consideration of overall level of functioning, length of time since the last episode of disturbance, total duration of the disturbance, and whether prophylactic treatment is being given.

0-Unspecified.

---

posed (117) that overlap DSM-III-R in many ways, but at the same time identify somewhat different populations of patients. This does not mean, however, that the situation is chaotic. Most patients identified by one set of criteria but not by another would be diagnosed as having "possible" or "suspected" schizophrenia by nearly all clinicians and investigators. Variations among different sets of criteria reflect differences in relative importance assigned to particular features with regard to differential diagnosis. In time further research should clarify these issues.

Natural History

As chronic schizophrenia typically begins insidiously, it is often hard to determine when the disorder started. In retrospect, the majority of patients show certain prepsychotic personality abnormalities: excessive shyness, social awkwardness, withdrawal from personal relationships, and inability to form close relationships (the so-called schizoid personality). A parent put it this way: "He was always afraid to make friends and felt that other people wouldn't like him. He couldn't be comfortable with girls, never knew what to say. I'd try to help him but it was very hard to change him. He'd cry and say it made him too uncomfortable to be with others." These traits may be present from early adolescence; often they are of concern to the patient's family for months or years before delusions or hallucinations become manifest. In addition, a number of studies have revealed an increased frequency of pregnancy and birth complications in the records of individuals who later developed schizophrenia (63) as well as in the children of schizophrenic mothers (148).

A number of studies have shown that in childhood and early adolescence, long before the diagnostic symptoms of schizophrenia are evident, schizophrenics have more academic difficulty in school and achieve lower scores on intelligence tests than do their siblings and other controls (1, 76). Instead of concluding that early academic difficulty and lowered intelligence may be early manifestations of schizophrenia, some investigators (65, 94) have argued that low IQ and schizophrenia are independently transmitted, but that a low IQ is one of the factors that increases the risk of clinical schizophrenia developing in those genetically predisposed. Other work indicates that antisocial and delinquent behavior may be an early manifestation of the illness (110), as may other personality disorders involving social withdrawal and disengagement (56, 97, 116, 143).

Delusions, hallucinations, and strange behavior usually start in the twenties. At first, these aberrations may be brief and vague, so that the family is not certain of their significance. Gradually they become more obvious and disturbing, usually leading to psychiatric consultation.

Schizophrenia infrequently begins after forty. The illness generally has a fluctuating course. One or more psychiatric hospitalizations commonly occur. Many schizophrenics spend most of their lives in psychiatric hospitals. Even when not hospitalizeed, schizophrenics lead disturbed lives: They usually fail to form satisfactory personal relationships, less often marry than their contemporaries, have poor job histories, and seldom achieve positions of responsibility. They may become neighborhood eccentrics or socially isolated residents of inner cities, doing irregular unskilled work or being supported by welfare. Many of the homeless people in large cities are probably schizophrenic.

Two developments have altered the general clinical course in these patients: the introduction of antipsychotics and the shift away from prolonged hospitalization. With antipsychotics, control of hallucinations, delusions, and bizarre behavior is possible in a substantial number of cases. As a result and because of the policy of early discharge, many patients spend far less time in psychiatric hospitals than was the case in earlier years, though rehospitalization for relatively brief periods is common.

The effectiveness of the antipsychotic agents, which seem nearly always to be associated with blocking of dopamine receptors in the brain, has led to the hypothesis that many of the clinical manifestations of schizophrenia depend on certain as yet unidentified dopaminergic neurons. This hypothesis has stimulated much research, but it must still be regarded as speculative (23, 24, 123). Studies of postmortem schizophrenic brains have been inconsistent in their findings on the levels of various biogenic amines, including dopamine, norepinephrine, and serotonin (11, 14, 27). On the other hand, several investigators have reported either increased binding to, or increased numbers of, dopamine receptors in postmortem schizophrenic brains. The influence of previous treatment with dopamine-receptor-blocking antipsychotic drugs on such findings is still uncertain, because animal studies show that prolonged treatment with these antipsychotic agents may lead to similar increases in brain dopamine receptors (78, 85, 124).

Good prognosis cases generally begin more abruptly than poor prognosis cases without a history of long-standing personality abnormalities. Although poor and good prognosis cases may differ somewhat clinically, the most important difference relates to outcome (35). When patients are seen for the first time, however, it may be difficult to predict what the course and prognosis will be. Several studies have identified criteria associated with the difference in prognosis among the schizophrenic disorders. Table 2.2 summarizes the data from these studies. When most of the criteria are present, the studies indicate that the prognosis associated with the criteria will be accurate in the majority of cases. A recent report (59) suggests that these criteria work better when the diagnosis of schizophrenia includes a broader range of disorders than are included in DSM-III-R. This may only indicate that the criteria help distinguish true schizophrenics from patients with other disorders—such as psychotic affective illness and acute intoxications, that can mimic schizophrenia.

Some patients presenting with an acute schizophrenic picture lose their psychotic features during hospitalization and begin to look like depressives (61, 88, 113). Others recover from the psychotic episode and then after a prolonged remission present with typical depression (118). These observations are of great theoretical interest. Some have interpreted them to suggest that the distinction between schizophrenic and affective disorders may be too arbitrary (118); others, emphasizing the familial prevalence of affective illness in the overlapping cases, have argued that such cases are more appropriately classified as affective disorders (139).

Complications

In no illness is it more difficult than in chronic schizophrenia to distinguish between the typical clinical picture and the complications. For example, difficulties in school or at work are very common features of schizophrenia. Yet it is usually very hard to tell whether these problems arise because of the patient's disordered thinking or loss of motivation or whether they result from the reactions of teachers, fellow students and workers, or supervisors of the patient's abnormal behavior. Obviously, the situation is one involving a complex matrix of forces in which it is impossible to decide what is the primary source of the difficulty. A suspicious, fearful, and deluded individual may not perform well in school or at work because of preoccupation with abnormal thoughts; at the same time, the withdrawn, preoccupied, and unresponsive stu-

Table 2.2 Nomenclature and prognostic criteria for good and poor prognosis cases

	Good Prognosis	Poor Prognosis
Diagnostic terms	Schizophreniform illness	Schizophrenia
	Acute Schizophrenia	Process Schizophrenia
	Schizoaffective Schizophrenia	Nuclear Schizophrenia
	Reactive Schizophrenia	Chronic Schizophrenia
	Remitting Schizophrenia	Nonremitting Schizophrenia
Prognostic criteria		
Mode of Onset	Acute	Insidious
Precipitating Events	Frequently reported	Usually not reported
Prepsychotic History	Good	Poor; frequent history of "schizoid" traits (aloofness, social isolation)
Confusion	Often present	Usually absent
Affective Symptoms	Often present and prominent	Often absent or minimal; affective responses usually "blunted" or "flat"
Marital Status	Usually married	Often single, especially males
Family History of Affective Disorder	Often present	May be present but less likely
Family History of Schizophrenia	Absent or rare	Increased

dent or worker may lead to criticism and other adverse responses from teachers and supervisors. These, in turn, will reinforce the individual's pathological responses. The definition of the disorder encompasses its natural history and at the same time specifies the complications. These include impaired education, poor work history and job achievement, celibacy, and prolonged psychiatric hospitalization. An increased risk of suicide among young patients may also be a complication.

Because chronic schizophrenia typically begins early in life and is characterized by recurrent or persistent manifestations, the patient's schooling and education suffer. Early school difficulties have been noted, but even among those who do well in elementary school, difficulties may arise in high school or college. Social withdrawal and loss of interest in studies may become evident. These changes coupled with the need to leave school at the onset of more dramatic symptoms eventually lead to dropping out of school in many cases. If the illness peaks after the completion of formal education, the same clinical features may lead to marked reduction of effectiveness at work, demotions, being fired, frequent job changes, and financial dependency.

In the past, early and prolonged hospitalization was associated with high rates of celibacy among schizophrenic patients. Schizophrenic men were particularly affected, presumably because male initiative was more important than female initiative in getting married. Thus a schizophrenic man would be less successful in courtship, whereas a schizophrenic woman might attract a suitor despite her illness. With the advent of modern drug treatment and the reduction in prolonged psychiatric hospitalization, marriage and childbearing rates of schizophrenics have approached those of the general population, though the sex difference persists (36, 130).

Despite the great reduction in chronic psychiatric hospitalization, no group of patients spends more time in psychiatric hospitals than schizophrenics. The majority are hospitalized for repeated, *relatively* brief periods, but a substantial minority of patients, perhaps one-quarter or one-third, spend many years in hospitals.

A common fear about deluded schizophrenic patients is that they are likely to act on their delusions and commit crimes. Available data suggest, however, that there is little or no increased risk of significant crimes committed by schizophrenics (54). They may be arrested for vagrancy, disturbing the peace, or similar misdemeanors, but only rarely are they involved in felonies.

Some authors have reported an increased suicide risk among young schizophrenics (30). Unfortunately, these authors did not attempt to distinguish between poor and good prognosis cases. It may be that the increased suicide risk is largely a function of the latter state; if so, this would suggest a link between at least some cases of good prognosis illness and primary affective disorders (108).

Family Studies

All investigators have found an increased prevalence of schizophrenia among the close relatives of schizophrenics (44, 50, 53, 67, 71, 72). Most studies have indicated a prevalence of between 10 and 15 percent among first-degree relatives of index cases compared with a general population figure of under 1 percent.

All but one of a series of twin studies (Table 2.3) have shown significantly higher concordance rates for schizophrenia among monozygotic twins than among dizygotic twins of the same sex (2, 37, 40, 49, 50, 62, 66, 75, 84, 111, 119, 135). The prevalence of schizophrenia among the latter is similar to the prevalence among ordinary siblings of the same sex. Although the actual concordance rates have varied from study to study, probably depending on methods of ascertainment (49, 50, 69), the rate for monozygotic twins is generally three to six times that of dizygotic twins.

The failure to find complete concordance in monozygotic twins has naturally led to the conclusion that there must be important environmental factors in the etiology of schizophrenia. Ofter such environmental factors are assumed to be of a social or psychological nature. One report (17), however, suggested that the discordance may be related to certain "brain development deviations"

Table 2.3 Schizophrenia concordance rates in twins

	MZ Twins		DZ Twins (same sex)	
	"Strict" Schizo- phrenia (%)	Including "Border- line" Cases (%)	"Strict" Schizo- phrenia (%)	Including "Border- line" Cases (%)
Investigator				
Luxemberger, 1928 (Germany)	50	71	0	0
Rosanoff et al., 1934 (USA)	44	61	9	13
Essen-Möller, 1941 (Sweden)	14	71	8	17
Kallmann, 1946 (USA)		69		17
Slater, 1953 (UK)		65		14
Inouye, 1963 (Japan)		60		18
Tienari, 1963 (Finland)	6	31	5	5
Kringlen, 1966 (Norway)	28	38	6	14
Gottesman and Shields, 1966 (UK)	42	54	9	18
Fischer, et al., 1969 (Denmark)	24	48	10	19
Allen et al., 1972 (USA)		27		5

Adapted from Fischer et al. (40) and Allen et al. (2)

to which monozygotic twins may be "specially prone." The report indicated that concordance for schizophrenia in monozygotic twins was close to 100 percent when both twins were right-handed but that it fell markedly when one or both of the twins was not clearly right-handed, suggesting that the risk of schizophrenia is somehow associated with the process of brain lateralization. Other reports have been contradictory (83, 134).

In addition, it should be noted that as discordant monozygotic twins are followed, many pairs become concordant (9), suggesting that the age of onset may be affected by nongenetic factors in some cases.

There have been two studies of children separated early in life from schizophrenic parents and raised by unrelated adoptive parents. In one of these studies, done in the United States, schizophrenia was found in five of forty-seven children of hospitalized schizophrenic mothers and no cases were found in fifty control children (60). All five schizophrenic children had been hospitalized, three chronically; the other two were taking antipsychotic drugs. In the other study, made in Denmark, about 32 percent of the children of schizophrenic parents received a diagnosis in the "schizophrenia spectrum," compared to about 18 percent of controls (112); none, however, had been hospitalized for schizophrenia. It is unclear how many would be considered schizophrenic in the "narrow" sense. "Schizophrenia spectrum" refers to a range of disorders, including schizophrenia and personality disorders that appear to cluster in certain families.

Another Danish study approached the problem of hereditary predisposition to schizophrenia in a different way. Monozygotic twin pairs in which one twin was schizophrenic, whereas the other was not were identified. The children of these discordant twins were studied. It was found that the frequency of schizophrenia was the same in children of nonaffected discordant schizophrenic twins as in children of affected members of discordant twin pairs (39). The prevalence of schizophrenia in all these children was about 10 percent, which is similar to the general figure for children of schizophrenic parents (49, 66).

The results of all these investigations point to the likelihood of a hereditary predisposition to schizophrenia.

The distinction between good and poor prognosis cases generally has not been studied in investigations of the familial transmission of schizophrenia. Available evidence suggests, however, that the prevalence of affective disorders is increased in close relatives of good prognosis cases. One interpretation given these data is that at least some good prognosis cases are atypical forms of affective illness (108).

The recognition that schizophrenic disorders are familial and that they probably result in part from hereditary factors has focused attention on identifying at an early age those individuals within a vulnerable family who are at greatest risk of developing the clinical disorders. Such "high-risk" studies are under way in many countries (93, 116, 132). One of the first studies, conducted in Denmark (116), indicates that the risk of schizophrenia during a ten-year follow-up of ten- to twenty-year-old children of schizophrenic mothers is about eight times greater than in matched control children. The risks of "borderline states (including schizoid and paranoid personality disorders)," psychiatric hospitalization, and suicide are also much greater in the children of schizophrenic mothers than in controls. The reported data further suggest that certain measurements of autonomic nervous system function may help identify those high-risk children who will develop the clinical disorder. The same group has found that childhood formal thought disorder correlates positively with formal thought disorder in adulthood, which may mean that schizophrenic symptoms usually develop gradually over many years (98).

A frequent problem in studies of the familial distribution of psychiatric disorders is the failure to examine both parents of index cases. As a result puzzling familial associations between different disorders may be observed. For example, in the adoption studies of schizophrenia described earlier, antisocial personality disorders have been found to be associated with schizopherenia. This may indicate that antisocial behavior can be one of the manifestations of schizophrenia, but an equally likely explanation is the tendency of some schizophrenics to mate with individuals manifesting other psychiatric disorders, especially antisocial personality (41, 42, 43, 96, 126).

Recent studies have also shown a familial and even a genetic relationship between schizophrenia and schizotypal personality, thus supporting the validity of the schizophrenia-spectrum concept (71, 72, 136, 137). Schizotypal personality, described in DSM-III-R, includes many of the chronic personality features seen in schizophrenia, but it does not include delusions or hallucinations.

Differential Diagnosis

The differential diagnosis between poor and good prognosis cases has already been discussed (see pp. 55, 56). Most studies indicate that when patients fulfill the criteria of either good prognosis or poor prognosis illness, in the majority of cases the clinical course and long-term prognosis will be consistent with the classification. Certain data suggest that the improved outcome in some series of poor prognosis schizophrenics may be the result of improved treatment (103). A major element in the differentiation between good and poor prognosis cases is "premorbid adjustment." Inconsistent findings may reflect difficulties in defining and measuring such adjustment (121, 131, 132), and additional studies are needed in which patients are clearly separated on the basis of how long they have been sick prior to psychiatric evaluation.

A small number of patients hospitalized for depression will, at follow-up, turn out to be suffering from chronic schizophrenia (109). This is much more likely to be the case when the depression is characterized by striking delusions or hallucinations. Some patients who present with the symptoms of hysteria (Briquet's syndrome) will show at follow-up the full clinical picture of chronic schizophrenia (147), but this also is rare.

Obsessional disorder, early in its course, occasionally is difficult to distinguish from schizophrenia. If the obsessions are bizarre or if the patient *clearly* does not have insight into the abnormal nature of his thoughts and impulses, it may not be possible to make a confident diagnosis. The risk of schizophrenia in patients with obsessional disorder is small after the first year or two of illness and when the obsessions are classical (48).

A clinical picture resembling schizophrenia has been described in association with temporal lobe epilepsy (28, 95, 120). Patients presenting this clinical picture cannot be distinguished from typical schizophrenics, but the absence of an increased prevalence of schizophrenia in their close relatives suggests that one is dealing with a separate entity.

Although many good prognosis cases are probably manifestations of affective illness, the question of differential diagnosis is often one of diagnostic conventions or style. Certain delusions (those of poverty, sinfulness, disease) are common in depressions, and others (those of overconfidence and unusual powers) in mania. Different delusions (those of control, persecution, depersonalization) are more likely to lead to a diagnosis schizophrenia.

The differential diagnosis of good prognosis illness and organic brain syndrome may also at times be a matter of convention. If disorientation and memory impairment are transient and less striking than the delusions, hallucinations, and bizarre behavior, a diagnosis of schizophrenia may be made. Patients with systemic medical illnesses affecting the brain such as lupus erythematosus may develop psychiatric symptoms that vary over time from those of depression to those of a brain syndrome, with a schizophrenic picture in between. Most patients with brain syndromes, however, present no problem in differential diagnosis because they experience little or nothing in the way of delusions or hallucinations; even when these symptoms are present, the persistent disorientation and memory impairment point to the correct diagnosis.

Persistent alcoholic hallucinosis is a syndrome of striking hallucinations, chiefly auditory, with a clear sensorium, that follows alcohol withdrawal. It may occur either after other withdrawal symptoms have subsided or in the absence of other manifestations of withdrawal. Alcoholic hallucinosis usually subsides within a couple of weeks. Occasionally, it persists and becomes chronic, thus resembling schizophrenia. Family studies have been undertaken to determine whether there is an increased prevalence of schizophrenia in close relatives (thus suggesting that the alcoholism precipitated or was superimposed on typical schizophrenia) or whether there is no increased prevalence (suggesting that, like epilepsy, alcoholism may produce a "symptomatic schizophrenia"). Unfortunately, the findings are contradictory and the issue cannot yet be resolved (115, 117).

Amphetamines, when taken chronically in large quantities, can cause a schizophrenialike illness (25). This psychotic state will almost always subside within ten days after the drug is discontinued. In rare chronic cases, the situation is similar to that seen in chronic alcoholic hallucinosis, where it is uncertain which came first—the drug abuse or the schizophrenic disorder.

The use of other psychotomimetic drugs such as LSD or PCP is responsible for a substantial percentage of cases of acute psychosis seen in emergency rooms or admitted to psychiatric hospitals. When first seen, it is not possible to distinguish such cases from naturally occurring acute schizophrenia, and sometimes one finds that a patient with chronic schizophrenia is suffering an acute intensification of the illness as the result of using these drugs.

Clinical Management

Antipsychotic medication, coupled with a policy of early discharge from psychiatric hospitals, made possible in part because of the antipsychotics, has become the keystone of standard treatment. Occasionally, the combination of antipsychotics and electroconvulsive treatments may produce better results than antipsychotics alone (122). Delusions, hallucinations, and bizarre behavior may be at least partially controlled by adequate doses of antipsychotics, though these drugs probably do not affect the emotional blunting and social withdrawal associated with schizophrenia. Larger doses are usually needed during acute exacerbations but may be reduced as the patient's symptoms abate.

The prolonged use of antipsychotics may lead to certain undesirable consequencees. Even at moderate dosage levels, parasympatholytic, or atropinelike, effects and Parkinsonism are common. At higher doses, lenticular opacities (34, 46, 101) and, more important, tardive dyskinesia (64, 68) have been described in many long-term patients. Patients receiving large doses of thioridazine may develop a condition resembling retinitis pigmentosa that causes visual impairment and blindess (55). A variety of skin rashes (150), a form of intrahepatic obstructive jaundice (which may rarely lead to chronic biliary cirrhosis) (73), and bone narrow toxicity may occur (89, 99), all probably as a result of special hypersensitivity. Despite these and other untoward reactions, the introduction of the antipsychotics represents a major advance in the treatment of schizophrenia.

The drugs do not control all schizophrenic symptoms. Patients continue to suffer from intermittent delusions and hallucinations, personality changes, and general social impairment. Many experienced clinicians believe, though, that the reduction in chronic hospitalization and the greater efforts at rehabilitation resulting from an increased optimism about more effective treatment prevent some personality deterioration.

The return of chronic schizophrenic patients to the community is not without problems. Many patients are severe burdens to their families. Although they may be better following drug treatment, they are not well. They continue to experience symptoms and their adjustment to family life is difficult and stressful. Also, most chronic schizophrenic patients are not able to support themselves financially and require family or community help. Recent work (79, 86, 125, 142) suggests that some schizophrenic patients do better and require fewer hospitalizations if their family members show less "expressed emotion," referring to overt criticism and pressure on the ill relative. In some cases, efforts to help the family with these reactions seem to result in milder symptoms in the patient and fewer hospitalizations.

The long-term care of chronic schizophrenics requires patience and perseverance. Many patients will omit medication if not carefully supervised; some require parenteral medication at least intermittently. Rehospitalization for brief periods may be necessary to reinstitute drug treatment or to help the family through a bad period in the course of the patient's illness. Many schizophrenic patients will respond to suggestions about their symptoms and adjustment; some are able to participate in more elaborate psychotherapeutic interviews. Nearly always, the patient's family will need support, encouragement, guidance, and understanding. Without this help, many families are unable to cope with the patient and all treatment then becomes more difficult.

Because good prognosis cases are generally episodic, they present different therapeutic problems. As in chronic schizophrenia, antipsychotics may help. In many cases, electrotherapy produces remission. The place of lithium in the treatment of some cases, particularly those with features of mania, is still unclear; some authors favor it, others do not.

Generally, after symptoms of acute schizophrenia have subsided, no further treatment is indicated unless there are complications such as alcoholism or drug abuse.

References

1. Albee, G., Lane, E., and Reuter, J. M.   Childhood intelligence of future schizophrenics and neighborhood peers. J. Psychol. 58:141–144, 1964.
2. Allen, M. G., Cohen, S., and Pollin, W.   Schizophrenia in veteran twins: a diagnostic review. Am. J. Psychiat. 128:939–945, 1972.
3. Andreasen, N. C., Dennert, J. W., Olsen, S. A., and Damasio, A. R. Hemispheric asymmetries and schizophrenia. Am. J. Psychiat. 139:427–430, 1982.
4. Andreasen, N., Nasrallah, H. A., Dunn, V., Olson, S. C., Grove, W. M., Ehrhardt, J. C., Coffman, J. A., and Crossett, J. H. W. Structural abnormalities in the frontal system in schizophrenia. Arch. Gen. Psychiat. 43:136–144, 1986.
5. Andreasen, N. C., Olsen, S. A., Dennert, J. W., and Smith, M. R. Ventricular enlargement in schizophrenia: relationship to positive and negative symptoms. Am. J. Psychiat. 139:297–302, 1982.
6. Andreasen, N. C., Smith, M. R., Jacoby, C. G., Dennert, J. W., and Olsen, S. A.   Ventricular enlargement in schizophrenia: definition and prevalence. Am. J. Psychiat. 139:292–296, 1982.
7. Astrup, C., Fossum, A., and Holmboe, R.   *Prognosis in Functional Psychoses.* Springfield, Ill.: C. C. Thomas, 1962.
8. Astrup, C., and Noreik, K.   *Functional Psychoses: Diagnostic and Prognostic Models.* Springfield, Ill.: C. C. Thomas, 1966.
9. Belmaker, R., Pollin, W., Wyatt, R. J., and Cohen, S.   A follow-up of monozygotic twins discordant for schizophrenia. Arch. Gen. Psychiat. 30:219–222, 1974.
10. Benes, F. M., Davidson, J., and Bird, E. D.   Quantitative cytoarchitectural studies of the cerebral cortex of schizophrenics. Arch. Gen. Psychiat. 43:31–35, 1986.
11. Bennett, J. P., Jr., Enna, S. J., Bylund, D. B., Gillin, J. C., Wyatt, R. J., and Snyder, S. H.   Neurotransmitter receptors in frontal cortex of schizophrenics. Arch. Gen. Psychiat. 36:927–934, 1979.
12. Berman, K. F., Zec, R. F., and Weinberger, D. R.   Physiologic dysfunc-

tion of dorsolateral prefrontal cortex in schizophrenia. Arch. Gen. Psychiat. 43:126–135, 1986.

13. Bigelow, L. B., Nasrallah, H. A., and Rauscher, F. P. Corpus callosum thickness in chronic schizophrenia. Brit. J. Psychiat. 142:284–287, 1983.

14. Bird, E. D., Spokes, E. G. S., and Iversen, L. L. Increased dopamine concentration in limbic areas of brain from patients dying with schizophrenia. Brain 102:347–360, 1979.

15. Bleuler, E. Dementia Praecox or the Group of Schizophrenias (trans. J. Zinkin). New York: International Universities Press, 1950.

16. Bogerts, B., Meertz, E., and Schonfeldt-Bausch, R. Basal ganglia and limbic system pathology in schizophrenia. Arch. Gen. Psychiat. 42:784–791, 1985.

17. Boklage, C. E. Schizophrenia, brain asymmetry development, and twinning: cellular relationship with etiological and possibly prognostic implications. Biol. Psychiat. 12:19–35, 1977.

18. Boyd, J. H., Pulver, A. E., and Stewart, W. Season of birth: schizophrenia and bipolar disorder. Schizophr. Bull. 12:173–186, 1986.

19. Brenner, D. Elementary Textbook of Psychoanalysis. New York: International Universities Press, 1955.

20. Brill, N., and Glass, J. Hebephrenic schizophrenic reactions. Arch. Gen. Psychiat. 12:545–551, 1965.

21. Brown, R., Colter, N., Corsellis, J. A. N., Crow, T. J., Frith, C. D., Jagoe, R., Johnstone, E. C., and Marsh, L. Postmortem evidence of structural brain changes in schizophrenia. Arch. Gen. Psychiat. 43:36–42, 1986.

22. Buchsbaum, M. S., Cappelletti, J., Ball, R., Hazlett, E., King, A. C., Johnson, J., Wu, J., and DeLisi, L. E. Positron emission tomographic image measurement in schizophrenia and affective disorders. Ann. Neurol. 15:S157–S165, 1984.

23. Carlsson, A. Does dopamine play a role in schizophrenia? Psychol. Med. 7:583–598, 1977.

24. Carlsson, A. Antipsychotic drugs, neurotransmitters, and schizophrenia. Am. J. Psychiat. 135:164–173, 1978.

25. Connell, P. H. Amphetamine Psychosis, Maudsley Monogr., no. 5. London: Oxford Univ. Press, 1958.

26. Crow, T. J. A re-evaluation of the viral hypothesis: is psychosis the result of retroviral integration at a site close to the cerebral dominance gene? Brit. J. Psychiat. 145:243–253, 1984.

27. Crow, T. J., Baker, H. F., Cross, A. J., Joseph, M. H., Lofthouse, R., Longden, A., Owen, F., Riley, G. J., Glover V., and Killpack, W. S. Monoamine mechanisms in chronic schizophrenia: post-mortem neurochemical findings. Brit. J. Psychiat. 134:249–256, 1979.

28. Davison, K. Schizophrenic-like psychoses associated with organic cerebral disorders: a review. Psychiat. Dev. 1:1–34, 1983.

29. Dennert, J. W., and Andreasen, N. C. CT scanning and schizophrenia: a review. Psychiat. Dev. 1:105–121, 1983.

30. Drake, R. E., Gates, C., Whitaker, A., and Cotton, P. G. Suicide among schizophrenics: a review. Compr. Psychiat. 26:90–100, 1985.
31. Dunham, H. W. *Community and Schizophrenia*. Detroit: Wayne State Univ. Press, 1965.
32. Early, T. S., Reiman, E. M., Raichle, M. E., and Spitznagel, E. L. Left globus pallidus abnormality in never-medicated patients with schizophrenia. Proc. Natl. Acad. Sci. USA 84:561–563, 1987.
33. Editorial. Hearing loss and perceptual dysfunction in schizophrenia. Lancet 2:848–849, 1981.
34. Edler, K., Gottfries, C. G., Haslund, J., and Raun, J. Eye changes in connection with neuroleptic treatment especially concerning phenothiazines and thioxanthenes. Acta Psychiat. Scand. 47:377–385, 1971.
35. Eitlinger, L., Laane, C. V., and Langfeldt, G. The prognostic value of the clinical picture and the therapeutic value of physical treatment in schizophrenia and the schizophreniform states. Acta Psychiat. et Neurol. Scand. 33:33–53, 1958.
36. Erlenmeyer-Kimling, L., Nicol, S., Rainer, J. D., and Deming, E. Changes in fertility rates of schizophrenic patients in New York State. Am. J. Psychiat. 125:916–927, 1969.
37. Essen-Möller, E. *Psychiatrische Untersuchungen an einer Serie von Zwillingen*. Copenhagen: Ejnar Munksgaard, 1941.
38. Ey, H. Unity and diversity of schizophrenia: clinical and logical analysis of the concept of schizophrenia. Am. J. Psychiat. 115:706–714, 1959.
39. Fischer, M. Psychoses in the offspring of schizophrenic twins and their normal co-twins. Brit. J. Psychiat. 118:43–52, 1971.
40. Fischer, M., Harvald, B., and Hauge, M. A Danish twin study of schizophrenia. Brit. J. Psychiat. 115:981–990, 1969.
41. Fowler, R. C., and Tsuang, M. T. Spouses of schizophrenics: a blind comparative study. Compr. Psychiat. 16:339–342, 1975.
42. Fowler, R. C., Tsuang, M. T., and Cadoret, R. J. Parental psychiatric illness associated with schizophrenia in the siblings of schizophrenics. Compr. Psychiat. 18:271–275, 1977.
43. Fowler, R. C., Tsuang, M. T., and Cadoret, R. J. Psychiatric illness in the offspring of schizophrenics. Compr. Psychiat. 18:127–134, 1977.
44. Frangos, E., Athanassenas, G., Tsitourides, S., Katsanou, N., and Alexandrakou, P. Prevalance of DSM III schizophrenics among the first-degree relatives of schizophrenic probands. Acta Psychiat. Scand. 72: 382–386, 1985.
45. Goldberg, E. M., and Morrison, S. L. Schizophrenia and social class. Brit. J. Psychiat. 109:785–802, 1963.
46. Gombos, G., and Yarden, P. E. Ocular and cutaneous side-effects after prolonged chlorpromazine treatment. Am. J. Psychiat. 123:872–874, 1967.
47. Goodwin, D. W., Alderson, P., and Rosenthal, R. Clinical significance of hallucinations in psychiatric disorders. Arch. Gen. Psychiat. 24:76–80, 1971.

48. Goodwin, D. W., Guze, S. B., and Robins, E. Follow-up studies in obsessional neurosis. Arch. Gen. Psychiat. 20:182–187, 1969.
49. Gottesman, I. I., and Shields, J. Contributions of twin studies to perspectives in schizophrenia. In *Progress in Experimental Personality Research*, Maher, B. A. (ed.), 3:1–84. New York: Academic Press, 1966.
50. Gottesman, I. I., and Shields, J. *Schizophrenia: The Epigenetic Puzzle*. Cambridge: Cambridge Univ. Press, 1982.
51. Gur, R. E., Resnick, S. M., Alavi, A., Gur, R. C., Caroff, S., Dann, R., Silver, F. L., Saykin, A. J., Chawluk, J. B., Kushner, M., and Reivich, M. Regional brain function in schizophrenia. I. A positron emission tomography study. Arch. Gen. Psychiat. 44:119–125, 1987.
52. Gur, R. E., Resnick, S. M., Gur, R. C., Alavi, A., Caroff, S., Kushner, M., and Reivich, M. Regional brain function in schizophrenia. II. Repeated evaluation with positron emission tomography. Arch. Gen. Psychiat. 44: 126–129, 1987.
53. Guze, S. B., Cloninger, C. R., Martin, R. L., and Clayton, P. J. A follow-up and family study of schizophrenia. Arch. Gen. Psychiat. 40: 1273–1276, 1983.
54. Guze, S. B., Goodwin, D. W., and Crane, J. B. Criminality and psychiatric disorders. Arch. Gen. Psychiat. 20:583–591, 1969.
55. Hagopian, V., Stratton, D., and Busiek, R. Five cases of pigmentary retinopathy associated with thioridazine administration. Am. J. Psychiat. 123:97–100, 1966.
56. Hanson, D. R., Gottesman, I. I., and Heston, L. L. Some possible childhood indicators of adult schizophrenia inferred from children of schizophrenics. Brit. J. Psychiat. 129:142–154, 1976.
57. Hare, E. H. Season of birth in schizophrenia and neurosis. Am. J. Psychiat. 132:1168–1171, 1975.
58. Hare, E., and Moran, P. A relation between seasonal temperature and the birth rate of schizophrenic patients. Acta Psychiat. Scand. 63:396–405, 1981.
59. Harrow, M., Westermeyer, J. F., Silverstein, M., Strauss, B. S., and Cohler, B. J. Predictors of outcome in schizophrenia: the process-reaction dimensions. Schizophr. Bull. 12:195–207, 1986.
60. Heston, L. Psychiatric disorders in foster home-reared children of schizophrenic mothers. Brit. J. Psychiat. 112:819–825, 1966.
61. Hoedemaker, F. S. Psychotic episodes and postpsychotic depression in young adults. Am. J. Psychiat. 127:606–610, 1970.
62. Inouye, E. Similarity and dissimilarity of schizophrenia in twins. Proceedings of the 3d World Congress on Psychiatry, 1:524–530. Montreal: Univ. of Toronto Press, 1963.
63. Jacobsen, B., and Kinney, D. K. Perinatal complications in adopted and non-adopted schizophrenics and their controls. Acta Psychiat. Scand. 62, Suppl. 285:337–346, 1980.
64. Jeste, D. V., and Wyatt, R. J. Therapeutic strategies against tardive dyskinesia. Arch. Gen. Psychiat. 39:803–815, 1982.

65. Jones, M. B., and Offord, D. R. Independent transmission of IQ and schizophrenia. Brit. J. Psychiat. 126:185–190, 1975.
66. Kallmann, F. J. The genetic theory of schizophrenia. An analysis of 691 twin index families. Am. J. Psychiat. 103:309–322, 1946.
67. Karlsson, J. L. Family transmission of schizophrenia: a review and synthesis. Brit. J. Psychiat. 140:600–606, 1982.
68. Kazamatsuri, H., Chien, C., and Cole, J. O. Therapeutic approaches to tardive dyskinesia. Arch. Gen. Psychiat. 27:491–499, 1972.
69. Kendler, K. S. Overview: a current perspective on twin studies of schizophrenia. Am. J. Psychiat. 140:1413–1425, 1983.
70. Kendler, K. S., Gruenberg, A. M., and Tsuang, M. T. Subtype stability in schizophrenia. Am. J. Psychiat. 142:827–832, 1985.
71. Kendler, K. S., Gruenberg, A. M., and Tsuang, M. T. Psychiatric illness in first-degree relatives of schizophrenic and surgical control patients. Arch. Gen. Psychiat. 42:770–779, 1985.
72. Kendler, K. S., Masterson, C. C., and Davis, K. L. Psychiatric illness in first-degree relatives of patients with paranoid psychosis, schizophrenia and medical illness. Brit. J. Psychiat. 147:524–531, 1985.
73. Kohn, N., and Myerson, R. Xanthomatous biliary cirrhosis following chlorpromazine. Am. J. Med. 31:665–670, 1961.
74. Kraepelin, E. Dementia Praecox and Paraphrenia (trans. R. M. Barclay; ed. G. M. Robertson). Edinburgh: E. & S. Livingstone, 1919.
75. Kringlen, E. Schizophrenia in twins. An epidemiological-clinical study. Psychiatry 29:172–184, 1966.
76. Lane, E., and Albee, G. W. Childhood intellectual differences between schizophrenic adults and their siblings. Am. J. Orthopsychiat. 35:747–753, 1965.
77. Langfeldt, G. The prognosis in schizophrenia. Acta Psychiat. et Neurol. Scand., Suppl. 110, 7–66, 1956.
78. Lee, T., and Seeman, P. Elevation of brain neuroleptic/dopamine receptors in schizophrenia. Am. J. Psychiat. 137:191–197, 1980.
79. Leff, J., Kuipers, L., Berkowitz, R., Vaughn, C., and Sturgeon, D. Life events, relatives' expressed emotion and maintenance neuroleptics in schizophrenic relapse. Psychol. Med. 13:799–806, 1983.
80. Leonhard, K. Aufteilung der endogenen Psychosen. Berlin: Akademie-Verlag, 1966.
81. Lippmann, S., Manshadi, M., Baldwin, H., Drasin, G., Rice, J., and Alrajeh, S. Cerebellar vermis dimensions on computerized tomographic scans of schizophrenic and bipolar patients. Am. J. Psychiat. 139:667–668, 1982.
82. Luchins, D. Computed tomography in schizophrenia. Arch. Gen. Psychiat. 39:859–860, 1982.
83. Luchins, D., Pollin, W., and Wyatt, R. J. Laterality in monozygotic schizophrenic twins: an alternative hypothesis. Biol. Psychiat. 15:87–93, 1980.
84. Luxenburger, H. Vorläufiger Bericht über psychiatrischen Serieunter-

suchungen an Zwillingen. Z. Ges. Neurol. Psychiat. 176:297–326, 1928.

85. Mackay, A. V. P., Bird, E. D., Spokes, E. G., Rossor, M., Iversen, L. L., Creese, I., and Snyder, S. H.   Dopamine receptors and schizophrenia: drug effect on illness. Lancet 2:915–916, 1980.

86. MacMillan, J. F., Gold, A., Crow, T. J., Johnson, A, L., and Johnstone, E. C.   Expressed emotion and relapse. Brit. J. Psychiat. 148:133–143, 1986.

87. Mathew, R. J., Meyer, J. S., Francis, D. J., Schoolar, J. C., Weinman, M., and Mortel, K. F.   Regional cerebral blood flow in schizophrenia: a preliminary report. Am. J. Psychiat. 138:112–113, 1981.

88. McGlashan, T. H., and Carpenter, W. T., Jr.   Postpsychotic depression in schizophrenia. Arch. Gen. Psychiat. 33:231–239, 1976.

89. McKinney, W., and Kane, F. J., Jr.   Pancytopenia due to chlorpromazine. Am. J. Psychiat. 123:879–880, 1967.

90. Merrin, E. L.   Schizophrenia and brain asymmetry. An evaluation of evidence for dominant lobe dysfunction. J. Nerv. Ment. Dis. 169:405–416, 1981.

91. Nasrallah, H. A., Jacoby, C. G., McCalley-Whitters, M., and Kuperman, S.   Cerebral ventricular enlargement in subtypes of schizophrenia. Arch. Gen. Psychiat. 39:774–777, 1982.

92. Nasrallah, H. A., Olson, S. C., McCalley-Whitters, M., Chapman, S., and Jacoby, C. G.   Cerebral ventricular enlargement in schizophrenia. Arch. Gen. Psychiat. 43:157–159, 1986.

93. Neuchterlein, K. H.   Childhood precursors of adult schizophrenia. J. Child Psychol. Psychiat. 27:133–144, 1986.

94. Offord, D. R.   School performance of adult schizophrenics, their siblings and age mates. Brit. J. Psychiat. 125:12–19, 1974.

95. Öhman, R., and Vogel, A. (eds.).   Temporal lobe epilepsy, psychotic and neurological manifestations, clinical tradition and new vistas. Acta Psychiat. Scand. 69, Suppl. 313, 1984.

96. Parnas, J.   Mates of schizophrenic mothers. Brit. J. Psychiat. 146:490–497, 1985.

97. Parnas, J., Schulsinger, F., Schulsinger, H., Mednick, S. A., and Teasdale, T. T.   Behavioral precursors of schizophrenia spectrum. Arch. Gen. Psychiat. 39:658–664, 1982.

98. Parnas, J., and Schulsinger, H.   Continuity of formal thought disorder from childhood to adulthood in a high-risk sample. Acta Psychiat. Scand. 74:246–251, 1986.

99. Pisciotta, A. V.   Agranulocytosis induced by certain phenothiazine derivatives. JAMA 208:1862–1868, 1969.

100. Planansky, K., and Johnston, R.   Depressive syndrome in schizophrenia. Acta Psychiat. Scand. 57:207–218, 1978.

101. Prien, R. F., DeLong, S. L., Cole, J. O., and Levine, J.   Ocular changes occurring with prolonged high dose chlorpromazine therapy. Arch. Gen. Psychiat. 23:464–468, 1970.

102. Retterstöl, N. *Paranoid and Paranoiac Psychoses.* Springfield, Ill.: C. C. Thomas, 1966.
103. Retterstöl, N. *Prognosis and Paranoid Psychoses.* Springfield, Ill.: C. C. Thomas, 1970.
104. Reveley, A. M., Reveley, M. A., and Murray, R. M. Cerebral ventricular enlargement in non-genetic schizophrenia: a controlled twin study. Brit. J. Psychiat. *144*:89–93, 1984.
105. Reveley, M. A. Ventricular enlargement in schizophrenia. The validity of computerised tomographic findings. Brit. J. Psychiat. *147*:233–240, 1985.
106. Reveley, M. A., Glover, V., Sandler, M., and Spokes, E. G. Brain monoamine oxidase activity in schizophrenics and controls. Arch. Gen. Psychiat. *38*:663–665, 1981.
107. Reynolds, G. P. Increased concentrations of lateral asymmetry of amygdala dopamine in schizophrenia. Nature *305*:527–529, 1983.
108. Robins, E., and Guze, S. B. Establishment of diagnostic validity in psychiatric illness: its application to schizophrenia. Am. J. Psychiat. *126*: 983–987, 1970.
109. Robins, E., and Guze, S. B. Classification of affective disorders: the primary-secondary, the endogenous-reactive, and the neurotic-psychotic concepts. In *Recent Advances in the Psychobiology of the Depressive Illnesses*, 283–293 Williams, T. A., Katz, M. M., and Shield, J. A., Jr. (eds.). Washington, D.C.: NIMH, DHEW, 1972.
110. Robins, L. *Deviant Children Grown Up.* Baltimore: Williams & Wilkins, 1966.
111. Rosanoff, A. J., Handy, I. M., Plesset, I. R., and Brush, S. The etiology of so-called schizophrenic psychoses. With special reference to their occurrence in twins. Am. J. Psychiat. *91*:247–286, 1934.
112. Rosenthal, D., Wender, P. H., Kety, S. S., Welner, J., and Schulsinger, F. The adopted-away offspring of schizophrenics. Am. J. Psychiat. *128*:307–311, 1971.
113. Roth, S. The seemingly ubiquitous depression following acute schizophrenic episodes, a neglected area of clinical discussion. Am. J. Psychiat. *127*:51–58, 1970.
114. Salokangas, R. R. K. Social class of the parents of schizophrenic patients. Proceedings of the 18th Nordic Psychiatric Congress. Acta Psychiat. Scand. 54, Suppl. 265:30, 1976.
115. Schuckit, M. A., and Winokur, G. Alcoholic hallucinosis and schizophrenia: a negative study. Brit. J. Psychiat. *119*:549–550, 1971.
116. Schulsinger, H. A ten-year follow-up of children of schizophrenic mothers. Clinical assessment. Acta Psychiat. Scand. *53*:371–386, 1976.
117. Scott, D. F. Alcoholic hallucinosis—an aetiological study. Brit. J. Addict. *62*:113–125, 1967.
118. Sheldrick, C., Jablensky, A., Sartorius, N., and Shepherd, M. Schizophrenia succeeded by affective illness: catamnestic study and statistical enquiry. Psychol. Med. *7*:619–624, 1977.

119. Slater, E. Psychotic and Neurotic Illnesses in Twins. London: HMSO, 1953.
120. Slater, E., Beard, A., and Glithero, E.   The schizophrenia-like psychoses of epilepsy. Brit. J. Psychiat. 109:95:–150, 1963.
121. Small, N. E., Mohs, R. C., Halperin, R., Rosen, W. G., Masterson, C., Kendler, K. S., Horvath, T. B., and Davis, K. L.   A study of the reliability of reported premorbid adjustment in schizophrenic patients. Biol. Psychiat. 19:203–211, 1984.
122. Smith, K.   ECT-chlorpromazine and chlorpromazine compared in the treatment of schizophrenia. J. Nerv. Ment. Dis. 144:284–290, 1967.
123. Snyder, S. H.   The dopamine hypothesis of schizophrenia: focus on the dopamine receptor. Am. J. Psychiat. 133:197–202, 1976.
124. Snyder, S. G.   Dopamine receptors, neuroleptics, and schizophrenia. Am. J. Psychiat. 138:460–464, 1981.
125. Spiegel, D., and Wissler, T.   Family environment as a predictor of psychiatric rehospitalization. Am. J. Psychiat. 143:56–60, 1986.
126. Stephens, D. A., Atkinson, M. W., Kay, D. W., Roth, M., and Garside, R. F.   Psychiatric morbidity in parents and sibs of schizophrenics and non-schizophrenics. Brit. J. Psychiat. 127:97–108, 1975.
127. Stephens, J. H.   Long-term course and prognosis of schizophrenia. Semin. Psychiat. 2:464–485, 1970.
128. Stephens, J. H., and Astrup, C.   Prognosis in "process" and "non-process" schizophrenia. Am. J. Psychiat. 119:945–953, 1963.
129. Stephens, J. H., Astrup, C., Carpenter, W. T., Jr., Shaffer, J. W., and Goldberg, J.   A comparison of nine systems to diagnose schizophrenia. Psychiat. Res. 6:127–143, 1982.
130. Stevens, B. C.   Marriage and Fertility of Women Suffering from Schizophrenia or Affective Disorders. Maudsley Monogr., no. 19. London: Oxford Univ. Press, 1969.
131. Strauss, J. S., and Gift, T. E.   Choosing an approach for diagnosing schizophrenia. Arch. Gen. Psychiat. 34:1248–1253, 1977.
132. Strauss, J. S., Kokes, R., Klorman, R., and Sacksteder, J.   Premorbid adjustment in schizophrenia: concepts, measures, and implications. Symp. Schizophr. Bull. 3:182–245, 1977.
133. Tanaka, Y., Hazama, H., Kawahara, R., and Kobayashi, K.   Computerized tomography of the brain in schizophrenic patients. Acta Psychiat. Scand. 63:191–197, 1981.
134. Taylor, P. J., Dalton, R., Fleminger, J. J., and Lishman, W. A.   Differences between two studies of hand preference in psychiatric patients. Brit. J. Psychiat. 140:166–173, 1982.
135. Tienari, P.   Psychiatric illnesses in identical twins. Acta Psychiat. Scand., 39, Suppl. 171:1–195, 1963.
136. Torgersen, S.   Genetic and nosological aspects of schizotypal and borderline personality disorders. Arch. Gen. Psychiat. 41:546–554, 1984.
137. Torgersen, S.   Relationship of schizotypal personality disorder to schizophrenia: genetics. Schizophr. Bull. 11:554–563, 1985.

138. Torrey, E. F., Torrey, B. B., and Peterson, M. R. Seasonality of schizophrenic births in the United States. Arch. Gen. Psychiat. 34:1065–1070, 1977.
139. Tsuang, M. T., Dempsey, G. M., Dvoredsky, A., and Struss, A. A family history study of schizo-affective disorder. Biol. Psychiat. 12:331–338, 1977.
140. Vaillant, G. E. Prospective prediction of schizophrenic remission. Arch. Gen. Psychiat. 11:509–518, 1964.
141. Vaillant, G. E. The prediction of recovery in schizophrenia. J. Nerv. Ment. Dis. 135:534–543, 1962.
142. Vaughn, C. E., Snyder, K. S., Jones, S., Freeman, W. B., and Falloon, I. R. H. Family factors in schizophrenic relapse. Arch. Gen. Psychiat. 41:1169–1177, 1984.
143. Watt, N. F. Patterns of childhood development in adult schizophrenics. Arch. Gen. Psychiat. 35:160–165, 1978.
144. Weinberger, D. R., Kleinman, J. E., Luchins, D. J., Bigelow, L. B., and Wyatt, R. J. Cerebellar pathology in schizophrenia: a controlled postmortem study. Am. J. Psychiat. 137:359–361, 1980.
145. Wender, P. H., Rosenthal, D., Kety, S. S., Schulsinger, F., and Welner, J. Social class and psychopathology in adoptees. Arch. Gen. Psychiat. 28:318–325, 1973.
146. Wong, D. F., Wagner, H. N., Jr., Tune, L. E., Dannals, R. F., Pearlson, G. D., Links, J. M., Tamminga, C. A., Broussolle, E. P., Ravert, H. T., Wilson, A. A., Toung, J. K. T., Malat, J., Williams, J. A., O'Tuama, L. A., Snyder, S. H., Kuhar, M. J., and Gjedde, A. Positron emission tomography reveals elevated D_2 dopamine receptors in drug-naive schizophrenics. Science 234:1558–1563, 1986.
147. Woodruff, R. A., Clayton, P. J., and Guze, S. B. Hysteria: studies of diagnosis, outcome, and prevalence. JAMA 215:425–428, 1971.
148. Wrede, G., Mednick, S. A., Huttenen, M. O., and Nilsson, C. G. Pregnancy and delivery complications in the births of an unselected series of Finnish children with schizophrenic mothers. Acta Psychiat. Scand. 62:369–381, 1980.
149. Wyatt, R. J., Potkin, S. G., and Murphy, D. L. Platelet monoamine oxidase activity in schizophrenia: a review of the data. Am. J. Psychiat. 136:377–385, 1979.
150. Zelickson, A. Skin changes and chlorpromazine. JAMA 198:341–344, 1966.

3. Panic Disorder
(Anxiety Neurosis)

Definition

Panic disorder is a chronic illness characterized by recurrent, acute anxiety attacks that have a definite onset and spontaneous termination. During attacks the patient is fearful and has symptoms associated with the autonomic nervous system: palpitations, tachycardia, rapid or shallow breathing, dizziness, and tremor. Between attacks, patients may be relatively asymptomatic though some experience fatigue, headache, and individual components of anxiety attacks in a persistent fashion. Panic disorder is not synonymous with anxiousness, which is a symptom rather than a syndrome. Anxiety symptoms and attacks can be part of the course of any psychiatric illness. When they occur in the absence of other significant psychiatric symptoms, the diagnosis of panic disorder is made.

Panic disorder has been known by many names (Table 3.1). Introduced by DSM-III, the term "panic disorder" replaced "anxiety neurosis" as the official name for the disorder. Panic disorder and anxiety neurosis will be used interchangeably in this chapter. Anxiety neurosis has a century-long history and is still used by many clinicians.

DSM-III-R introduced the concept that panic disorder should be subdivided into two types: panic disorder with agoraphobia and panic disorder without agoraphobia. A third category, agoraphobia without panic attacks, is described in chapter 6. The criteria for panic disorder are provided in Table 3.2. Criteria for agoraphobia are in Table 6.3.

DSM-III also introduced the term "generalized anxiety disorder" to refer to individuals who have persistent feelings of anxiety

Table 3.1 Terms used to describe anxiety neurosis

Neurasthenia
Neurocirculatory asthenia
Nervous exhaustion
DaCosta's syndrome
Effort syndrome
Irritable heart
Soldier's heart
Somatization psychogenic cardiovascular reaction
Somatization psychogenic asthenic reaction
Anxiety reaction
Vasomotor neurosis
Personalities with mixed psychic and physical anomalies:
  constitutionally labile with tendency to functional dis-
  orders of specific organ systems.

From Cohen and White (13)

but not panic attacks or phobias. Criteria for generalized anxiety disorder are presented in Table 3.4. Because this is a relatively new category and has been little studied, this chapter is primarily devoted to panic disorder.

Anxiety is often distinguished from ordinary fear by its lack of an appropriate stimulus to explain the emotion. To highlight the apparently unmotivated nature of anxiety, the term "free-floating" anxiety is often used.

Sometimes, however, patients do experience anxiety attacks in response to a fear-provoking situation such as facing an angry employer or giving a public speech. In these cases the clinician must decide whether the anxiety is grossly out of proportion to the fear-provoking stimulus, as well as make a diagnosis based on the overall history.

Most anxiety neurotics report that they experience some anxiety attacks without any fear-provoking stimulus, but on other occasions they overreact to situations that would produce some degree of apprehension in individuals without the disorder.

Some clinicians separate state anxiety from trait anxiety. The former refers to anxiety at any particular moment: "I feel anxious right now." The latter refers to a tendency to be anxious over a long period: "I generally feel anxious." The two forms of anxiety commonly occur together, with increased life stresses raising the anxiety of an already anxiety-prone individual beyond the point of tolerance (51).

Historical Background

> O lift me from the grass!
> I die! I faint! I fail!
> . . .
> My cheek is cold and white, alas!
> My heart beats loud and fast.
>
> PERCY BYSSHE SHELLEY, *The Indian Serenade*

It has been suggested that Shelley was having an anxiety attack when he wrote these lines. If so, he probably would have called it something else. A good possibility is "vapors," a favorite expression in the nineteenth century for anxiety reactions that also referred to fainting. Anxiety neurotics sometimes faint—probably from hyperventilating—and in the nineteenth century fainting among women was fashionable. In Victorian times the prototype of a refined young woman was a "swooner, pale and trembling, who responded to unpleasant or unusual social situations by taking to the floor in a graceful and delicious maneuver, in no way resembling the crash of the epileptic" (19). A Jane Austen heroine found one social situation "too pathetic for the feelings of Sophie and myself. We fainted alternately on a sofa." Overtight corsets may have been responsible for some of the fainting. A nineteenth-century physician, Dr. John Brown, cured fainting by "cutting the stay laces, which ran before the knife and cracked like a bow string" (19).

One of the first medical terms for anxiety neurosis was neurasthenia, defined by an American physician, G. M. Beard, in 1869 (4). The term was used rather broadly to include patients with hysteria and obsessional illness as well as hypochondriacs, anxiety neurotics, and swooners (8).

In 1871 DaCosta reported a syndrome among Civil War military personnel that he called "irritable heart." His concept was probably more restricted than Beard's and closer to our definition of anxiety neurosis (18). DaCosta noted that for more than a century previously there had been similar reports of cases from British and German military medical sources.

The term "anxiety neurosis" was first used by Freud in 1895 to refer to patients who probably would have been called neurasthenic by other physicians at that time.

In 1950 Wheeler et al. published a twenty-year follow-up study of 173 patients with anxiety neurosis diagnosed by specific checklist criteria (76). Later Cohen et al. reported a family study based on a similar group of anxiety neurotics (12). These studies are the basis of the clinical concept of the disorder discussed in this chapter.

In the hundred years since Beard's paper on neurasthenia was published, there have been many theories regarding the cause of the disorder, ranging from constitutional weakness of the nervous system, through social and psychological factors, to more recent biochemical hypotheses. Although the cause of anxiety neurosis remains unknown, some studies have indicated that there are physiologic differences between anxiety neurotics and normal individuals (13, 37, 40, 41). Anxiety neurotics, for instance, are more responsive to painful stimuli of various types. They also have low exercise tolerance, and high blood levels of lactic acid following exercise. The latter observation has led to a theory that abnormal lactate levels may be responsible for producing anxiety symptoms.

Although hypocalcemia may not be the explanation for lactate-induced panic attacks, the fact that lactate can precipitate panic attacks in predisposed individuals is not disputed. Liebowitz et al. (44, 46) reviewed six studies and found that panic attacks followed lactate infusion in 70 to 100 percent of patients with panic disorder. They then found that lactate-induced panic attacks could be prevented by the antidepressant imipramine, which blocks spontaneous panic attacks. Pitts and McClure (58) suggested that panic was induced by direct chemical action of the lactate, but alternative explanations have been proposed, including the possibility that lactate functions as a conditioned stimulus producing a phobic response (44, 46).

Several studies (22, 24, 73, 79) have shown that panic symptoms can also be produced in panic-disorder patients by giving them carbon dioxide by inhalation. Propranolol, a drug used to relieve anxiety, reduces the symptoms produced by carbon dioxide.

Epidemiology

Anxiety neurosis is one of the most common psychiatric syndromes. A Boston study indicated that approximately 5 percent of the adult population is affected (13). Many patients, however, experience the syndrome in a mild form and probably never seek medical care for their symptoms (78). Many others consult family practitioners or internists rather than psychiatrists. Anxiety neurotics who see psychiatrists may represent a small group with a high prevalence of secondary affective disorder (78).

For these reasons, prevalence estimates of anxiety neurosis vary widely. The Framingham investigators (34) reported a rate of 11 percent in a sample of 960 persons. Weissman et al. (74), studying 1095 households in New Haven, Connecticut, found panic disorder in only 1.5 percent of the subjects. A national survey of psychotropic drug use reported a similar prevalence rate (72).

Compounding the problem of estimating prevalence is the recent tendency to consider agoraphobia a form of panic disorder. Many, if not most, agoraphobics have panic attacks, but until recently it was believed that panic disorder is considerably more common than agoraphobia. There is now some question about this. In the National Institute of Mental Health (NIMH) Environmental Catchment Area study, data were obtained from three sites totaling 9453 persons (63). The rates for panic disorder were almost identical in the three cities (New Haven, Baltimore, and St. Louis). The overall prevalence rate was 1.4 percent. When combined with agoraphobia, the prevalence ranged from 3.7 to 9.7 percent, with a combined prevalence of 6.1 percent. Conceivably, the earlier studies failed to ask whether agoraphobic symptoms were present or absent, thus the original report (13) that about 5 percent of adults had panic disorder may have been close to reality.

Women have panic disorder at least twice as often as men. In the Framingham study (34), 16 percent of women were affected compared to about 5 percent of men. In most family studies, women have outnumbered men about two to one (12, 16, 56, 60).

Anxiety neurotics do not differ from the general population in educational level or socioeconomic status (73). There is no evidence that any specific type of childhood experience such as bereavement or birth order predisposes to the disorder (13, 77).

Clinical Picture

Anxiety attacks are the hallmark of panic disorder (Table 3.2). The attacks usually begin suddenly, sometimes in a public place, sometimes at home, perhaps awakening the patient from sleep. There is a sense of foreboding, fear, and apprehension; a sense that one had suddenly become seriously ill; a feeling that one's life may be threatened by such illness. Among some patients there is a disturbing sense that one's body has changed or become distorted (depersonalization). Such a feeling of alien change may extend to the surrounding world (derealization).

Symptoms of labored breathing, smothering, palpitation, blurred vision, tremulousness, and weakness usually accompany the apprehension and foreboding. If a patient is examined during such an attack, signs of distress will be present: tachycardia, sweating, tachypnea, tremor, hyperactive deep tendon reflexes, and dilated pupils. An electrocardiogram taken during such an episode usually reveals sinus tachycardia. There may be localized or diffuse areas of tenderness over the anterior chest wall, possibly related to fatigue of the muscles from exaggerated thoracic breathing and involvement of costal cartilages and costovertebral joints (3).

Anxiety attacks vary in frequency among patients. Some experience them on a daily basis, others have them only once or twice

Table 3.2 Diagnostic criteria for Panic Disorder (DSM-III-R)

A. At some time during the disturbance, one or more panic attacks
(discrete periods of intense fear or discomfort) occurred that were
(1) unexpected, i.e., did not occur immediately before or on exposure to a situation that almost always caused anxiety, and (2)
not triggered by situations in which the person was the focus of
others' attention.

B. Either four attacks, as defined in criterion A, have occurred within
a four-week period, or one or more attacks have been followed by
a period of at least a month of persistent fear of having another
attack.

C. At least four of the following symptoms developed during at least
one of the attacks:
    (1)   shortness of breath (dyspnea) or smothering sensations
    (2)   dizziness, unsteady feelings, or faintness
    (3)   palpitations or accelerated heart rate (tachycardia)
    (4)   trembling or shaking
    (5)   sweating
    (6)   choking
    (7)   nausea or abdominal distress
    (8)   depersonalization or derealization
    (9)   numbness or tingling sensations (paresthesias)
    (10) flushes (hot flashes) or chills
    (11) chest pain or discomfort
    (12) fear of dying
    (13) fear of going crazy or of doing something uncontrolled
    Note: Attacks involving four or more symptoms are panic attacks;
    attacks involving fewer than four symptoms are "limited symp-
    tom attacks" [see chapter 6].

D. During at least some of the attacks, at least four of the C symptoms developed suddenly and increased in intensity within ten
minutes of the beginning of the first C symptom noticed in the
attack.

E. It cannot be established that an organic factor initiated and maintained the disturbance, e.g., Amphetamine or Caffeine Intoxication, hyperthyroidism

Note: Mitral valve prolapse may be an associated condition, but does
not preclude a diagnosis of Panic Disorder.

DSM-III-R describes two types of panic disorder: Panic Disorder with Agoraphobia and Panic Disorder without Agoraphobia. The criteria for Agoraphobia
are in Table 6.3. The three categories have separate codes.

Table 3.3 Symptoms of Anxiety Neurosis

Symptoms	60 Patients (%)	102 Controls* (%)
Palpitation	97	9
Tires easily	93	19
Breathlessness	90	13
Nervousness	88	27
Chest pain	85	10
Sighing	79	16
Dizziness	78	16
Faintness	70	12
Apprehensiveness	61	3
Headache	58	26
Paresthesias	58	7
Weakness	56	3
Trembling	54	17
Breathing unsatisfactory	53	4
Insomnia	53	4
Unhappiness	50	2
Shakiness	47	16
Fatigued all the time	45	6
Sweating	42	33
Fear of death	42	2
Smothering	40	3
Syncope	37	11
Nervous chill	24	0
Urinary frequency	18	2
Vomiting and diarrhea	14	0
Anorexia	12	3
Paralysis	0	0
Blindness	0	0

* Healthy controls consisted of 50 men and 11 women from a large industrial plant and 41 healthy postpartum women from the Boston Lying-In Hospital. From Cohen and White (13)

a year. Other symptoms may occur between attacks. Table 3.3 lists these symptoms and their frequency reported in one study. Again, the symptoms can occur in practically any pattern. If they occur frequently in *the absence of panic attacks*, the patient may have a condition that fits the DSM-III-R criteria for Generalized Anxiety Disorder (Table 3.4).

Table 3.4 Diagnostic criteria for Generalized Anxiety Disorder (DSM-III-R)

A. Unrealistic or excessive anxiety and worry (apprehensive expectation) about two or more life circumstances, e.g., worry about possible misfortune to one's child (who is in no danger) and worry about finances (for no good reason), for a period of six months or longer, during which the person has been bothered more days than not by these concerns. In children and adolescents, this may take the form of anxiety and worry about academic, athletic, and social performance.

B. If another . . . disorder is present, the focus of the anxiety and worry in A is unrelated to it, e.g., the anxiety or worry is not about having a panic attack (as in Panic Disorder), being embarrassed in public (as in Social Phobia), being contaminated (as in Obsessive Compulsive Disorder), or gaining weight (as in Anorexia Nervosa).

C. The disturbance does not occur only during the course of a Mood Disorder or a psychotic disorder.

D. At least 6 of the following 18 symptoms are often present when anxious (do not include symptoms present only during panic attacks):

Motor tension:

(1) trembling, twitching, or feeling shaky
(2) muscle tension, aches, or soreness
(3) restlessness
(4) easy fatigability

Autonomic hyperactivity:

(5) shortness of breath or smothering sensations
(6) palpitations or accelerated heart rate (tachycardia)
(7) sweating, or cold clammy hands
(8) dry mouth
(9) dizziness or light-headedness
(10) nausea, diarrhea, or other abdominal distress
(11) flushes (hot flashes) or chills
(12) frequent urination
(13) trouble swallowing or "lump in throat"

Vigilance and scanning

(14) feeling keyed up or on edge
(15) exaggerated startle response

(16) difficulty concentrating or "mind going blank" because of anxiety

(17) trouble falling or staying asleep

(18) irritability

E. It cannot be established that an organic factor initiated and maintained the disturbance, e.g., hyperthyroidism, caffeine intoxication.

Cardiorespiratory symptoms are the most frequent chief complaints that anxiety neurotics report to physicians: "I have heart spells," "I think I'll smother," or "There is no way for me to get enough air." The chief complaint is occasionally psychological, but more often it indicates that the patient considers his or her disorder medical, frequently with the fear that the disturbance may be very serious.

A fifty-one-year-old electrical engineer was brought to an emergency room by ambulance complaining of severe left anterior chest pain. His breathing was labored and he complained of numbness and tingling in his lips and fingers. His pulse was 110. He was perspiring heavily and obviously frightened.

The electrocardiogram was normal. Blood enzymes were within normal limits. Nevertheless, the patient was admitted to the intensive care unit. He was observed for twenty-four hours and then discharged asymptomatic.

The history revealed that the patient had first experienced severe chest pain at the age of twenty-one while watching a movie. He had many subsequent episodes of chest pain accompanied by dyspnea and anxiety and had been taken to an emergency room on at least ten previous occasions. Although the examinations were always normal, he had undergone a variety of cardiovascular procedures, including cardiac catheterization. He had never been seen by a psychiatrist and at no time was told that he very likely suffered from a common disorder called anxiety neurosis.

Sympoms may become associated with specific situations that patients will try to avoid. For example, they may choose aisle seats in theaters, preferably close to exits, so that if an attack occurs, they will not be confined. Or, patients may avoid social situations in which an attack would be both frightening and embarrassing.

Phobias are common in patients with anxiety neurosis. Some clinicians, in fact, do not distinguish between anxiety neurosis and phobic disorders. For our purposes, anxiety neurosis is diagnosed when anxiety symptoms predominate, phobic disorder when phobias predominate.

Natural History

Panic disorder begins early in life. Cohen and White (13) found the mean age of onset to be twenty-five. The disorder appears in most patients between the ages of eighteen and thirty-five. This is consistent with a more recent study (16) in which the mean age of onset was twenty-six years with a standard deviation of ten years. Some patients may remember the exact time and circumstances of the first attack. In a study by Winokur and Holemon, thirteen of thirty-one anxiety neurotics were able to describe such circumstances (77). These researchers found that four subjects remembered specifically that they had been awakened at night by their first anxiety attack. Some patients remember having their first attack at times of stress (e.g., while making a speech in class). Thus the disorder may begin acutely with a discrete anxiety attack, but it also may begin insidiously with feelings of tenseness, nervousness, fatigue, or dizziness for years before the first anxiety attack.

Patients' initial medical contact is not always helpful. If they come to a physician complaining of cardiorespiratory symptoms, fearful of heart disease, a physician unacquainted with the natural history of anxiety neurosis may support these patients' fears by referring them to a specialist and admonishing that exercise should be limited.

Although patients commonly present with cardiorespiratory symptoms, they do not do so invariably. Some patients present with symptoms of "irritable colon" (47). In fact, among patients with irritable colon, anxiety neurosis is one of the most common psychiatric illnesses found. Such patients usually consult a gastro-enterologist. Among their most common presenting symptoms are abdominal cramping, diarrhea, constipation, nausea, belching, fla-tus, and occasionally dysphagia. ("Irritable colon" sometimes oc-curs in the course of other psychiatric illnesses such as hysteria or primary affective disorder. It also occurs in the absence of psychi-atric illness. One study [33], in fact, suggests that food intolerance is often a major factor in the pathogenesis of the irritable bowel syndrome.)

In one long-term study of patients meeting the criteria for anx-iety neurosis, gastrointestinal and musculoskeletal symptoms were as frequent as cardiovascular ones (54). For example, abdominal pain or discomfort was not far behind chest pain or discomfort and muscular aching or tension in occurrence. And during anxiety attacks, noncardiovascular symptoms such as headache or abdomi-nal pain often became the primary focus of concern. A shift of symptoms over the follow-up period reflected the increasing chro-nicity of the illness. In this series, the more acute, dramatic, and disruptive manifestations tended to drop out over time. Anxiety attacks after five years were occurring in only half of the patients, and insomnia, nausea, vomiting, light-headedness, and fainting were greatly reduced.

Anxiety neurosis can occasionally be severe, but in the majority of cases the course is mild (76). Symptoms wax and wane in an irregular pattern that may or may not be associated with events and circumstances interpreted by the patient as stressful. Despite their symptoms, most anxiety neurotics live productively without social impairment. There is evidence that patients of lower social class have more severe symptoms and greater social impairment than patients with higher incomes and more education (56). One possible explanation is that lower-class people are less likely than more affluent people to encounter a physician who takes the time to provide reassurance by explaining the nature of the illness.

Seven long-term studies of patients with anxiety neurosis con-ducted in Boston, Oslo, Zurich, England and Iowa show remark-ably similar findings (51). On five- to twenty-year follow-up, about

Table 3.5 Disability from anxiety neurosis—twenty-year follow-up

Symptoms and Disability	60 Patients (%)	22 Men (%)	38 Women (%)
Well	12	13	11
Symptoms, no disability	35	46	29
Symptoms, mild disability	38	32	42
Symptoms, moderate or severe disability	15	9	18

From Wheeler et al. (76)

50 to 60 percent of patients had recovered or were much improved. About one in five patients continued to have moderate to severe disability (Table 3.5).

A follow-up of anxiety neurotics seen by a psychiatric consultation service revealed almost identical findings, suggesting that those relatively few patients who see a psychiatrist have a prognosis as favorable as the majority who do not (54). In this study, men of age thirty or more at the time of psychiatric consultation showed the poorest outcome. As the authors note, this may reflect a tendency for men to ignore lesser degrees of stress and seek medical care only when more severely impaired.

Most studies do not associate anxiety neurosis with reduced longevity. Two recent studies from the University of Iowa, however, tend to contradict this assumption (14, 15). In both, men with panic disorder were twice as likely to die as expected, largely from cardiovascular disease and suicide. Although the risk of suicide among patients with anxiety disorder is not known, several follow-up studies have found suicides occur in such patients (38, 65, 76). The finding in Iowa of excess cardiovascular deaths among panic disorder patients has not been reported before.

Complications

The complications of anxiety neurosis are less severe than those of psychiatric disorders such as affective disorder or alcoholism, in which judgment may be impaired and suicide is a risk. Judgment is not impaired by anxiety neurosis and suicide rarely occurs.

The anxiety neurotics seen by psychiatrists are probably a select group with an illness frequently complicated by either secondary affective disorder or by alcoholism (78).

Depression is the most common complication of anxiety neurosis. Woodruff and associates (78) found that 50 percent of anxiety neurotics seen in a psychiatric clinic gave a history of secondary depression. Of 112 patients with anxiety neurosis examined in the medical clinics of a general hospital, 44 percent reported episodes of depression compared with 7 percent of surgical controls (11). Although the symptoms of secondary and primary depression may be identical, the depressions seen in anxiety neurosis are relatively mild. In one study, two-thirds of the depressive episodes were of three-months duration or less (11). An equal number were associated with precipitating life events such as the loss of a family member through death or divorce, an association less commonly seen in primary affective disorder.

Because of the frequent occurrence of depression in anxiety neurosis, there has been speculation that anxiety neurosis is a variant of primary affective disorder. Both conditions are relieved by tricyclic antidepressant medication. Recent evidence indicates that there is an overlap of depression and anxiety neurosis in the relatives of depressed patients. Arguments against the illnesses being identical are: (a) primary affective disorder is associated with a high rate of suicide and anxiety neurosis is not; (b) anxiety neurosis has an earlier age of onset than primary affective disorder; (c) panic attacks can usually be precipitated in people with a history of spontaneous panic attacks by the intravenous infusion of sodium lactate, and this rarely happens in depressed controls (58); (d) failure of suppression in the dexamethasone suppression test occurs in about half of primary affective disorder patients but is rare in individuals with spontaneous panic attacks (5).

Drug abuse may occur as a complication of anxiety neurosis, the occasional result of overzealous medication by well-meaning physicians.

Family Studies

Panic disorder is a strikingly familial condition. Crowe (17) found that two-thirds of cases had relatives affected with the same condition and that the risk in first-degree relatives was about three to four times the rate for the general population. Although some family studies suggest an overlap in the transmission of panic disorder and depression and a common diathesis hypothesis has been proposed, depression is more frequent in the family of depressives as is panic disorder in the family of those with panic disorder.

Twin studies of anxiety disorders document a 30 to 40 percent concordance among monozygotic twins, contrasting with a 4 percent concordance among dyzygotic twins. This of course supports a genetic predisposition (68, 70, 71).

Harris et al. (28) examined the relationship between panic disorder and agoraphobia in the family members of patients with each diagnosis. The families of panic disorder patients had a morbidity risk of 20.5 percent for panic disorder but only 1.9 percent for agoraphobia. The families of agoraphobics, by contrast, had a morbidity rate of 7.7 percent for panic disorder and 8.6 percent for agoraphobia. The rate of the two diagnoses in controls was 4 percent for each condition. This study cast some doubt on the current tendency to view most cases of agoraphobia as a subtype of panic disorder because the latter ran true to type in families, overlapping very little with agoraphobia. There is, however, general agreement that most agoraphobics experience episodes of panic early in the course of their agoraphobic illness.

Several studies (12, 59) indicate an increased frequency of alcoholism among fathers of anxiety neurotics. However, it is not clear which came first: the anxiety attacks or the alcoholism. Alcoholics commonly experience anxiety attacks, particularly when hung over. Because anxiety neurosis and alcoholism both begin in adolescence or early adulthood, it is usually not possible to tell which developed first.

Differential Diagnosis

Anxiety symptoms and anxiety attacks can be part of any psychiatric illness. Anxiety symptoms often appear in primary affective disorder, obsessional illness, phobic disorder, hysteria, and alcoholism. In part, the diagnosis is based on chronology. Anxiety neurosis is diagnosed only if there are no other symptoms or if anxiety symptoms antedated the others.

Medical illnesses that produce symptoms resembling those of anxiety neurosis include cardiac arrhythmias (especially paroxysmal atrial tachycardia), angina pectoris, hyperthyroidism, pheochromocytoma, parathyroid disease, and mitral valve prolapse. Of these, mitral valve prolapse has received by far the greatest attention in recent years.

Mitral valve prolapse occurs in about 5 percent of the population (20, 50). Abnormal electrocardiograms (EKGs) are usually present, together with systolic clicks and late systolic murmurs, but there are usually no symptoms. Confirmation of the diagnosis is often made with the echocardiogram. Studies have shown that a high proportion of anxiety neurotics—32 to 50 percent—have mitral valve prolapse (29, 39). According to one study (35), it was also common in agoraphobia, an illness often accompanied by panic attacks.

The relationship between mitral valve prolapse and anxiety neurosis is not clear. In one study (23), the presence or absence of a prolapsed mitral valve had no influence on the induction of a panic attack by sodium lactate. In another study (39), anxiety symptoms in patients with prolapsed mitral valves responded as well to imipramine as did anxiety symptoms in patients with a normal valve.

Leaving aside this problem, the diagnosis of anxiety neurosis obviously requires the exclusion of medical conditions that give rise to similar symptoms. Clinicians who are not familiar with anxiety neurosis as a specific diagnostic entity may mistakenly attribute the symptoms to a physical condition. Studies show that a false positive diagnosis is made in 25 percent of patients considered to have angina pectoris, some of whom almost certainly have anxiety neurosis (2). The percentage of false positives actually may be higher because of disagreement about interpreting EKGs. In one study, three cardiologists interviewed fifty-seven patients with chest pain. If any of them diagnosed angina, there was only a 55 percent chance that his two colleagues would agree with him (64). In another study of 110 EKGs read by at least one of four cardiologists as compatible with arteriosclerotic heart disease, two of the readers agreed in 60 percent of the cases, three out of the four agreed in 40 percent, and all four agreed in less than 20 percent (31). Nearly half of patients undergoing postmortem examinations who had received a diagnosis of angina pectoris show no coronary artery narrowing (43). False positives and false negatives are even found in double-blind evaluations of the two-step exercise EKG (21).

Clearly, it is important to exclude medical illness before diagnosing anxiety neurosis, but because of diagreement among clinicians, this may not be as simple as it seems.

In any case, there appears to be no characteristic EKG abnormality associated with anxiety neurosis. This conclusion came from a Framingham study of 203 cases of anxiety neurosis and 757 control subjects free of cardiovascular disease (34).

The mistake probably made most often in the evaluation of anxiety symptoms is that of overlooking primary affective disorder. Among other factors, a family history of affective disorder should alert one to the possibility that a patient with anxiety symptoms may have primary affective disorder. Also, anxiety that begins for the first time after the age of forty is commonly part of a depressive syndrome rather than a manifestation of anxiety neurosis.

Hysteria involves a broader range of medical complaints with frequent hospitalizations and operations. Anxiety neurotics are hospitalized and operated on no more frequently than other individuals. Furthermore, anxiety neurosis is *not* characterized by dramatic and medically unexplained complaints in nearly *every* area of the review of symptoms, as is the case in hysteria. Finally, sexual and menstrual complaints and conversion symptoms (unexplained neurological symptoms), common in hysteria, do not appear frequently among patients with anxiety neurosis.

Obsessional illness and phobic disorder can be mistaken for anxiety nurosis. Phobic disorder is so similar that the distinction depends on whether or not phobias dominate the clinical picture. Obsessive compulsive disorder involves obsessive ruminations and compulsive rituals. Although these may occur in mild form among patients with anxiety neurosis, they are not characteristically severe.

Anxiety neurosis also should be distinguished from the so-called postaccident anxiety syndrome. In one series (52), patients who had experienced a life-threatening accident subsequently experienced a long period of free-floating anxiety, muscular tension, irritability, impaired concentration, repetitive nightmares reproducing the accident, and social withdrawal. These symptoms persisted for six months to three years after the accident. None of the patients had cardiorespiratory or other autonomic symptoms, the hallmark of a typical anxiety neurosis. Also, in some cases referred by courts, the possibility of monetary compensation may have been a contributing factor. Supporting this possibility are a number of studies—starting with one by Heim in 1892 of survivors of falls in the Alps (30)—reporting that most individuals do not have frightening dreams or anxiety after their accidents. In any event, there is no evidence that life-threatening events precipitate anxiety neurosis as defined in this chapter (55).

DSM-III describes a condition called "post-traumatic stress disorder, delayed type." This has the same clinical features as the postaccident anxiety syndrome, but the symptoms do not appear until at least six months after the trauma. In 1980 the Veterans Administration announced that post-traumatic stress disorder, delayed type, was a compensable disorder. This meant that for the first time since World War I, the Department of Veterans Benefits could consider disorders to be service connected when the symptoms appeared long after military discharge.

Many veterans responded by filing claims based on their belief that they suffered from post-traumatic stress disorder, delayed type, related to traumatic war experiences. The symptoms of the disorder were well publicized in the media and in brochures distributed by national service organizations. Rarely before had so many claimants presented themselves to psychiatric examiners having read printed symptom checklists describing the diagnostic features of the disorder for which they sought compensation. Some psychiatrists were offended by the "mechanical litany of complaints recited by a well-read claimant" (1). The decision to award compensation was made even more difficult by the almost total lack of evidence that "post-traumatic stress disorder, delayed type" exists as a clinical entity.

Anxiety attacks are occasionally accompanied by hyperventilation, which produces symptoms that aggravate those of anxiety neurosis. Such symptoms are caused by lowering of pCO_2 as well as other chemical alterations, including an elevation of lactic acid. Five deep breaths produced by yawning or sighing may be enough to alter pCO_2 and produce the characteristic symptoms (10): cerebral hypoxia, a slowing of the electroencephalogram (EEG), and respiratory alkalosis, which in turn can induce tetany.

It appears that there is a group of patients who have hyperventilation syndrome (HVS) but do not meet the criteria for anxiety neurosis. In contrast to studies of anxiety neurosis, it is not known whether HVS represents a discrete syndrome or is symptomatic of other psychiatric or medical conditions. Apparently, some individuals are more susceptible than others to the effects of hyperventilation.

In one study at the Yale–New Haven Hospital, the diagnosis of HVS was made on the basis of the patient's response to overbreathing (10). Each patient suspected of having HVS was asked to overbreathe for up to three minutes or until he or she became dizzy. If the patient's reported symptoms were reproduced in their entirety and if no other explanation for the symptoms could be adduced from physical examination, medical history, or laboratory tests, the diagnosis of HVS was considered established. However, many patients had a history of conversion reactions, hypochondriasis, or "psychosomatic" illnesses; most were young women. Others had organic diseases such as adrenal insufficiency and peptic ulcer. Thus, psychiatric conditions such as Briquet's syndrome (hysteria) or medical illness actually may have been the primary disorder, with hyperventilation occurring secondarily, as it sometimes does in anxiety neurosis. In short, whether or not HVS exists as a separate syndrome is still uncertain.

Finally, the differential diagnosis of panic disorder is not complete without considering the Chinese restaurant syndrome, apparently caused by monosodium glutamate in Chinese food. This syndrome is characterized by palpitations, numbness, paresthesias, generalized weakness and other symptoms that can mimic a panic disorder (80).

How far to pursue a physical explanation for symptoms that may represent an anxiety neurosis depends on clinical judgment, but a good deal of money is often invested in tests and consultations. In one study (54), anxiety neurotics received an average of five laboratory tests and two consultations with specialists. As the authors explained, "This is not unusual, since most patients did not consider their illness as psychiatric but saw themselves as persons suffering from undiagnosed physical conditions. This almost inevitably led to a search for an organic basis for the complaints. Sometimes the search was conducted against the physicians' better judgment in order that the patient might accept psychiatric referral. Even at that, the patients as a group were demanding and not easily satisfied. One patient's remark typified the attitude of many, "I do not want to be told what I haven't got. I want the doctor to find something." Another patient stated, "I know I am nervous, I have always been nervous, but don't try to tell me that is what is wrong."

Clinical Management

Psychiatric referral is rarely necessary for anxiety neurosis if the patient's physician has diagnosed the disorder correctly, understands its natural history, and is willing to discuss the syndrome with the patient. Telling the patient that there is nothing wrong physically is usually not enough. Some patients resent the implication that their symptoms might be psychological. Others continue to believe that they have a serious illness. The physician should agree that something is wrong and should describe the syndrome in lay terms. Many patients are relieved by such an explanation and thereby become receptive to further reassurance.

As noted earlier, some patients with irritable bowel syndrome or musculoskeletal complaints meet the criteria for anxiety neurosis. In these cases, treatment should focus primarily on the basic disorder—the anxiety neurosis—so that the presenting gastrointestinal or musculoskeletal complaints are viewed in this context and treated as conservatively as possible.

Supportive psychotherapy is almost always indicated, yet there is no evidence that psychological management more extensive than reassurance has any better effect than reassurance alone (76). Prolonged and expensive forms of psychotherapy are rarely indicated.

In the 1960s, Klein and colleagues in New York (39) and Marks and colleagues in England (51) found that antidepressants prevented panic attacks. The New York group focused on imipramine and the British group on monoamine oxidase inhibitors (MAOIs). Both types of drugs apparently had similar efficacy in preventing panic attacks, although there was some disagreement about the relief persisting after the drugs were discontinued and about the role of supportive or behavioral therapy in augmenting the effects of the drugs. In any case, antidepressants have become standard treatment for preventing panic attacks.

Except for the atypical benzodiazepine alprazolam, none of the benzodiazepines seems to prevent panic attacks. However, benzodiazepines reduce anticipatory anxiety between attacks. Many patients are given both drugs: a tricyclic antidepressant, for example, at night, and benzodiazepines during the day on an as-needed basis. Of the tricyclics, imipramine has been the most widely studied, but recent reports indicate that desipramine (48) and the nontricyclic "new generation" antidepressant trazodone are also effective (53). In recent studies, alprazolam, the atypical benzodiazepine, has been proved to be effective not only as an anxiolytic drug but also as an agent that prevents panic attacks (when administered in rather largish amounts) (9, 66, 67).

There has been much speculation about the action of these drugs in reducing anxiety and preventing panic attacks. One of the most promising leads is evidence that imipramine prevents panic attacks by decreasing norepinephrine turnover in the brain, whereas alprazolam affects both benzodiazepine receptors and noradrenergic systems (6, 7, 32). Much attention has been focused on the locus ceruleus, a nucleus that serves as a relay center for the putative alarm-fear-anxiety system. Drugs that increase the firing rate of locus ceruleus neurons increase anxiety levels, whereas drugs that block the firing of these neurons apparently have opposite effects. The septohippocampal system has also been associated with anxiety (26), although this remains conjectural. However, evidence that panic disorder involves changes in brain activity has come from studies using positron emission tomography (PET) (61, 62). Patients with panic disorder who are vulnerable to lactate-induced anxiety attacks have strikingly abnormal hemispheric asymmetries of parahippocampal blood flow, blood volume, and oxygen metabolism.

The beta-adrenergic blocking agent propranolol, a drug used for cardiac arrhythmias, also may be effective in the treatment of anxiety neurosis (25, 75). Propranolol has potentially serious side effects but not usually in the low doses effective in reducing anxiety symptoms (69). The only significant one reported has been increased wheezing in asthmatic patients. Although the efficacy of propranolol in treating anxiety neurosis has not definitely been established, there is little question that it blocks some of the physiologic symptoms.

References

1. Atkinson, R. M., Henderson, R. G., Sparr, L. F., and Deale, S. Assessment of Viet Nam veterans for posttraumatic stress disorder in Veterans Administration disability claims. Am. J. Psychiat. 139:1118–1121, 1982.

2. Banks, T., and Shugoll, G. Confirmatory physical findings in angina pectoris. JAMA 200:107–112, 1967.

3. Bass, C., Wade, C., Gardner, W. N., Cawley, R., Ryan, K. C., Hutchison, D. C. S.. and Jackson, G. Lancet, March 19, 1983.

4. Beard, G. M. Neurasthenia or nervous exhaustion. Boston Med. Surg. J. 3:217, 1869.

5. Carroll, B. J. The dexamethasone suppression test for melancholia. Brit. J. Psychiat. 140:292–304, 1982.

6. Charney, D. S., and Heninger, G. R. Noradrenergic function and the mechanism of action of antianxiety treatment. I. The effect of long-term alprazolam treatment. Arch. Gen. Psychiat. 42:458–467, 1985.

7. Charney, D. S., and Heninger, G. R. Noradrenergic function and the mechanism of action of antianxiety treatment. II. The effect of long-term imipramine treatment. Arch. Gen. Psychiat. 42:473–481, 1985.

8. Chatel, J. C., and Peele, R. A centennial review of neurasthenia. J. Psychiat. 126:1404–1413, 1970.

9. Chouinard, G., Annable, L., Fontaine, R., and Solyom, L. Alprazolam in the treatment of generalized anxiety and panic disorders: a double-blind placebo-controlled study. Psychopharmacol. 77:229–233, 1982.

10. Christensen, B. Studies on hyperventilation. II: Electrocardiographic changes in normal man during voluntary hyperventilation. J. Clin. Invest. 24:880, 1946.

11. Clancy, J., Noyes, R., Hoenk, P. R., and Slymen, D. J. Secondary depression in anxiety neurosis. J. Nerv. Ment. Dis. 166:846–850, 1978.

12. Cohen, M. E., Badal, D. W., Kilpatrick, A., Ried, E. W., and White, P. D. The high familial prevalence of neurocirculatory asthenia (anxiety neurosis, effort syndrome). Am. J. Human Genet. 3:126, 1951.

13. Cohen, M., and White, P. Life situations, emotions, and neurocirculatory asthenia (anxiety neurosis, neurasthenia, effort syndrome). Ass. Res. Nerv. Dis. Proc. 29:832–869, 1950.

14. Coryell, W., Noyes, R., and Clancy, J. Excess mortality in panic disorder: a comparison with primary unipolar depression. Arch. Gen. Psychiat. 39:701–703, 1982.

15. Coryell, W., Noyes, R., and House, J. D. Am. J. Psychiat. 143:508–510, 1986.

16. Crowe, R. R., Noyes, R., Pauls, D. L., and Slymen, D. A family study of panic disorder. Arch. Gen. Psychiat. 40:1065–1069, 1983.

17. Crowe, R. R. The genetics of panic disorder and agoraphobia. Psychiat. Dev. 2:171–185, 1985.

18. DaCosta, J. M. On irritable heart, a clinical form of functional cardiac disorder and its consequences. Am. J. Med. Sci. 61:17, 1871.

19. Dalessio, D. J. Hyperventilation: the vapors, effort syndrome, neurasthenia. JAMA 239:1401–1402, 1978.

20. Devereux, R. B., Perloff, J. K., Reichek, N., and Josephson, M. E. Mitral valve prolapse. Circulation 54, 1976.

21. Friedberg, C. K. The two step exercise electrocardiogram: a double blind

evaluation of its use in the diagnosis of angina pectoris. Circulation 26: 1254–1260, 1962.

22. Fryer, M. R., Uy, J., Martinez, B. S., Goetz, R., Klein, D. F., Fryer, A., Liebowitz, M. R., and Gorman, J. CO_2 challenge of patients with panic disorder. Am. J. Psychiat. 144:1080–1082, 1987.

23. Gorman, J. M., Fyer, A. F., Gliklich, J., King, D., and Klein, D. F. Effect of sodium lactate on patients with panic disorder and mitral valve prolapse. Am. J. Psychiat. 138:247–249, 1981.

24. Gorman, J. M., Askanazi, J., Liebowitz, M. R., Fryer, A. J., Stein, J., Kinney, J. M., and Klein, D. F. Response to hyperventilation in a group of patients with panic disorder. Am. J. Psychiat. 141:857–861, 1984.

25. Granville-Grossman, K. L., and Turner, P. The effect of propranolol on anxiety. Lancet 1:788–790, 1966.

26. Gray, J. A. The Neuropsychology of Anxiety. Oxford Univ. Press, 1982.

27. Grosz, H. J., and Farmer, B. B. Pitts and McClure's lactate-anxiety study revisited. Brit. J. Psychiat. 120:415–418, 1972.

28. Harris, E. L., Noyes, R., Crowe, R. R., and Chaudhry, D. R. A family study of agoraphobia: report of a pilot study. Arch. Gen. Psychiat. 40: 1061–1064, 1983.

29. Hartman, N., Kramer, R., Brown, W. T., and Devereux, R. B. Panic disorder in patients with mitral valve prolapse. Am. J. Psychiat. 139:669–670, 1982.

30. Heim, A. Remarks on Fatal Falls. Yearbook of the Swiss Alpine Club 27:327–337, 1892. Trans. R. Noyes and R. Kletti in Omega 3:45–52, 1972.

31. Higgins, I. T. Ischemic heart disease: the problem. Milbank Mem. Fund. Q. 43:23–31, 1965.

32. Iversen, S. Benzodiazepine Divided. Trimble, M. R. (ed.), 176–185. New York: John Wiley & Sons, 1983.

33. Jones, V. A., Shorthouse, M., McLaughlan, P., Workman, E., and Hunter, J. O. Food intolerance: a major factor in the pathogenesis of irritable bowel syndrome. Lancet 1982.

34. Kannel, W. B., Dawber, T. R., and Cohen, M. E. The electrocardiogram in neurocirculatory asthenia (anxiety neurosis or neurasthenia): a study of 203 neurocirculatory asthenia patients and 757 healthy controls in the Framingham study. Ann. Intern. Med. 49:1351–1360, 1958.

35. Kantor, J. S., Zitrin, C. M., and Zeldis, S. M. Mitral valve prolapse syndrome in agoraphobic patients. Am. J. Psychiat. 137:467–469, 1980.

36. Kelly, D., Mittchell-Heggs, N., and Sherman, D. Anxiety and the effects of sodium lactate assessed clinically and physiologically. Brit. J. Psychiat. 119:129–141, 1971.

37. Kelly, D., and Walter, C. J. S. A clinical and physiological relationship between anxiety and depression. Brit. J. Psychiat. 115:401–406, 1969.

38. Kerr, T. A., Schapira, K., and Roth, M. The relationship between premature death and affective disorders. Brit. J. Psychiat. 115:1277–1282, 1969.

39. Klein, D. F.   Importance of psychiatric diagnosis in prediction of clinical drug effects. Arch. Gen. Psychiat. 37:63–72, 1980.
40. Levander-Lindgren, M.   Studies in neurocirculatory asthenia (DaCosta's syndrome). Acta Med. Scand. 172:665–683, 1962.
41. Levander-Lindgren, M.   Studies in neurocirculatory asthenia (DaCosta's syndrome). Acta Med. Scand. 173:631–637, 1963.
42. Liberthson, R., Sheehan, D. V., King, M. E., and Weyman, A. E.   The prevalence of mitral valve prolapse in patients with panic disorders. Am. J. Psychiat. 143:511–515, 1986.
43. Liebow, I. M., and Oseasohn, R.   Relationship between selected clinical and electrocardiographic findings and post mortem lesions associated with ischemic heart disease. J. Chronic Dis. 17:609–617, 1964.
44. Liebowitz, M. R., Fryer, A. J., Gorman, J. M., Dillon, D., Appleby, I. L., Levy, G., Anderson, S., Levitt, M., Palij, M., Davies, S. O., and Klein, D. F.   Lactate provocation of panic attacks. Arch. Gen. Psychiat. 41: 764–770, 1984.
45. Liebowitz, M. R., Fryer, A. J., Gorman, J. M., Dillon, D., Davies, S., Stein, J. M., Cohen, B. S., and Klein, D. F.   Specificity of lactate infusions in social phobia versus panic disorders. Am. J. Psychiat. 142:947–950, 1985.
46. Liebowitz, M. R., Gorman, J. M., Fryer, A. J., Levitt, M., Dillon, D., Levy, G., Appleby, I. L., Anderson, S., Palij, M., Davies, S. O., and Klein, D. F.   Lactate provocation of panic attacks. II. Biochemical and physiological findings. Arch. Gen. Psychiat. 42:709–719, 1985.
47. Liss, J., Alpers, D., and Woodruff, R. A.   The "irritable colon" syndrome and psychiatric illness. Dis. Nerv. Syst. 34:151–157, 1973.
48. Lydiard, R. B.   Preliminary results of an open, fixed-dose study of desipramine in panic disorder. Psychopharmacol. Bull. 23:139–140, 1987.
49. Margraf, J., Eilers, A., and Roth, W. T.   Sodium lactate infusions and panic attacks: a review and critique. Psychosom. Med. 48:23–51, 1986.
50. Markiewicz, W., Stoner, J., London, E., Hunt, S. A., and Popp, R. L.   Mitral valve prolapse in one hundred presumably healthy young females. Circulation 53:464, 1976.
51. Marks, I., and Lader, M.   Anxiety states (anxiety neurosis): a review. J. Nerv. Ment. Dis. 156:3–18, 1973.
52. Modlin, H. C.   Postaccident anxiety syndrome: psychosocial aspects. Am. J. Psychiat. 123:1008–1012, 1967.
53. Mavissakalian, M., Perel, J., Bowler, K., and Dealy, R.   Trazodone in the treatment of panic disorder and agoraphobia with panic attacks. Am. J. Psychiat. 144:785–787, 1987.
54. Noyes, R., and Clancy, J.   Anxiety neurosis: a 5-year follow-up. J. Nerv. Ment. Dis. 162:200–205, 1976.
55. Noyes, R., and Kletti, R.   Depersonalization in the face of life-threatening danger: a description. Psychiatry 39:19–27, 1976.
56. Noyes, R., Clancy, J., Crowe, R., Hoenk, P. R., and Slymen, D. J.   The familial prevalence of anxiety neurosis. Arch. Gen. Psychiat. 35:1057–1059, 1978.

57. Noyes, R., Clarkson, C., Crowe, R. R., Yates, W. R., and McChesney, C. M. A family study of generalized anxiety disorder. Am. J. Psychiat. 144:1019–1024, 1987.
58. Pitts, F. N., and McClure, J. N. Lactate metabolism in anxiety neurosis. N. Eng. J. Med. 277:1329–1336, 1967.
59. Pitts, F. N., Meyer, J., Brooks, M., and Winokur, G. Adult psychiatric illness assessed for childhood parental loss, and psychiatric illness in family members: a study of 748 patients and 250 controls. Am. J. Psychiat. 121, Suppl. i–x, 1965.
60. Reich, J. The epidemiology of anxiety. J. Nerv. Ment. Dis. 174:129–136, 1986.
61. Reiman, E. M., Raichle, M. E., Butler, F. K., Herscovitch, P., and Robins, E. A focal brain abnormality in panic disorder, a severe form of anxiety. Nature 310:683–685, 1984.
62. Reiman, E. M. The study of panic disorder using positron emission tomography. Psychiat. Dev. 1:63–78, 1987.
63. Robins, L. N., Helzer, G. E., and Weissman, M. M. Lifetime prevalence of specific psychiatric disorders in three sites. Arch. Gen. Psychiat. 41:949–958, 1984.
64. Rose, G. A. Ischemic heart disease: chest pain questionnaire. Milbank Mem. Fund. Q. 43:32–39, 1965.
65. Rosenberg, C. M. Complications of obsessional neurosis. Brit. J. Psychiat. 114:477–478, 1968.
66. Sheehan, D. V., Claycomb, J. B., and Surman, O. S. The relative efficacy of alprazolam, phenelzine and imipramine in treating panic attacks and phobias. Paper presented at the 137th Annual Meeting of the American Psychiatric Association, Los Angeles, 1984.
67. Sheehan, D. V., Coleman, J. H., and Greenblatt, D. J. Some biochemical correlates of panic attacks with agoraphobia and their response to a new treatment. J. Clin. Psychopharmacol. 4:66–75, 1984.
68. Slater, B., and Shields, J. Genetical aspects of anxiety. Brit. J. Psychiat. Special Publication no, 3, Studies of anxiety, 62–71. Ashford, Kent: Headley Bros., 1969.
69. Tanna, V. T., Penningrowth, R. P., and Woolson, R. F. Propranolol in anxiety neurosis. Compr. Psychiat. 18:319–326, 1977.
70. Torgersen, S. Genetics of neurosis: the effects of sampling variation upon the twin concordance ratio. Brit. J. Psychiat. 142:126–132, 1983.
71. Torgersen, S. Genetic factors in anxiety disorders. Arch. Gen. Psychiat. 40:1085–1089, 1983.
72. Uhlenhuth, E. H., Balter, M. B., Mellinger, G. D., Cisin, I. H., and Clinthorne, J. Symptom check list syndromes in the general population: correlations with psychotherapeutic drug use. Arch. Gen. Psychiat. 40:1167–1173, 1983.
73. Van Den Hout, M. A., and Griez, E. Panic symptoms after inhalation of carbon dioxide. Brit. J. Psychiat. 144:503–507, 1984.
74. Weissman, M. M., Meyers, J. K., and Harding, P. S. Psychiatric disor-

ders in a U.S. urban community: 1975–76. Am. J. Psychiatry 135:459–462, 1978.

75. Wheatley, D. Comparative effects of propranolol and chlordiazepoxide in anxiety states. Brit. J. Psychiat. 115:1411–1412, 1969.

76. Wheeler, E. O., White, P. D., Ried, E. W., and Cohen, M. E. Neurocirculatory asthenia (anxiety neurosis, effort syndrome, neuroasthenia). JAMA 142:878–889, 1950.

77. Winokur, G., and Holemon, E. Chronic anxiety neurosis: clinical and sexual aspects. Acta Psychiat. Scand. 39:384–412, 1963.

78. Woodruff, R. A., Jr., Guze, S. B., and Clayton, P. J. Anxiety neurosis among psychiatric outpatients. Compr. Psychiat. 13:165–170, 1972.

79. Woods, S. W., Charney, D. S., Loke, J., Goodman, W. K., Redmond, D. E., and Heninger, G. R. Carbon dioxide sensitivity in panic anxiety. Arch. Gen. Psychiat. 43:900–907, 1986.

80. Zautcke, J. L., Schwartz, J. A., and Mueller, E. J. Chinese restaurant syndrome: a review. Ann. Emergency Med. 15:1210–1213, 1986.

4. Hysteria
(Somatization Disorder)

Definition

The terms "hysteria" and "conversion symptoms" are both frequently used and, for some, they are synonymous. It is important, however, to distinguish between them.

Hysteria is a *polysymptomatic* disorder that begins early in life (usually in the teens, rarely after the twenties), chiefly affects women, and is characterized by recurrent, multiple somatic complaints often described dramatically. Characteristic features, all unexplained by other known clinical disorders, are varied pains, anxiety symptoms, gastrointestinal disturbances, urinary symptoms, menstrual difficulties, sexual and marital maladjustment, nervousness, mood disturbances, and conversion symptoms. Repeated visits to physicians and clinics, the use of a large number of medications—often at the same time—prescribed by different physicians, and frequent hospitalizations and operations result in a florid medical history (28, 29). Somatization Disorder is the term for hysteria in DSM-III-R (Table 4.1). Briquet's syndrome is another term for the same condition.

It should be emphasized, however, that the diagnostic criteria for somatization disorder are less stringent than those for Briquet's syndrome and were proposed to offer a short-cut to the diagnosis. As yet, there are few data about the validity of somatization disorder derived from systematic controlled follow-up and family studies; the findings thus far indicate that the two sets of criteria select overlapping but different populations (14, 20).

Conversion symptoms are unexplained symptoms suggesting neurological disease such as amnesia, unconsciousness, paralysis, "spells," aphonia, urinary retention, difficulty in walking, anesthe-

Table 4.1 Diagnostic criteria for Somatization Disorder (DSM-III-R)

A. A history of many physical complaints or a belief that one is sickly, beginning before the age of 30 and persisting for several years.

B. At least 13 symptoms from the list below. To count a symptom as significant, the following criteria must be met:

    (1) no organic pathology or pathophysiologic mechanism (e.g., a physical disorder or the effects of injury, medication, drugs, or alcohol) to account for the symptom or, when there is related organic pathology, the complaint or resulting social or occupational impairment is grossly in excess of what would be expected from the physical findings

    (2) has not occurred only during a panic attack

    (3) has caused the person to take medicine (other than over-the-counter pain medication), see a doctor, or alter lifestyle

Symptom list:

Gastrointestinal symptoms:

    (1) **vomiting (other than during pregnancy)**

    (2) abdominal pain (other than when menstruating)

    (3) nausea (other than motion sickness)

    (4) bloating (gassy)

    (5) diarrhea

    (6) intolerance of (gets sick from) several different foods

Pain symptoms:

    (7) **pain in extremities**

    (8) back pain

    (9) joint pain

    (10) pain during urination

    (11) other pain (excluding headaches)

Cardiopulmonary symptoms:

    (12) **shortness of breath when not exerting oneself**

    (13) palpitations

    (14) chest pain

    (15) dizziness

Conversion or pseudoneurologic symptoms:

    (16) **amnesia**

    (17) **difficulty swallowing**

(18) loss of voice
(19) deafness
(20) double vision
(21) blurred vision
(22) blindness
(23) fainting or loss of consciousness
(24) seizure or convulsion
(25) trouble walking
(26) paralysis or muscle weakness
(27) urinary retention or difficulty urinating

Sexual symptoms for the major part of the person's life after opportunities for sexual activity:

(28) **burning sensation in sexual organs or rectum (other than during intercourse)**
(29) sexual indifference
(30) pain during intercourse
(31) impotence

Female reproductive symptoms judged by the person to occur more frequently or severely than in most women:

(32) **painful menstruation**
(33) irregular menstrual periods
(34) excessive menstrual bleeding
(35) vomiting throughout pregnancy

Note: The seven items in boldface may be used to screen for the disorder. The presence of two or more of these items suggests a high likelihood of the disorder.

---

sia, and blindness—the so-called pseudoneurological or grand hysterical symptoms (39). "Unexplained" means only that the history, neurological examination, and diagnostic tests have failed to reveal a satisfactory explanation for the symptoms. The term "conversion symptom" has no etiologic or pathogenetic implication; it refers, in a descriptive way only, to a limited group of symptoms. In order to give the term greater specificity, unexplained pains and other unexplained medical symptoms that do not suggest neurological disease are not included in the definition. If other unexplained medical symptoms such as headaches, backaches, and abdominal pains were included, conversion symptom would mean *any* unexplained medical symptom and thus would lose any precision it has (28, 29).

Sometimes a related syndrome occurs as a sudden, short-lived epidemic (mass hyteria), characteristically in a school population and most frequently affecting female students (40, 53). Blackouts, dizziness, weakness, headaches, hyperventilation, nausea, and abdominal pain are the usual symptoms. Suggestion and fear are thought to be major factors in the pathogenesis of these episodes. Long-term follow-up of the affected children has not been reported, though the short-term prognosis appears to be very good.

In summary, conversion symptoms comprise a limited group of individual symptoms suggesting neurological disease. Hysteria is a polysymptomatic syndrome that typically includes conversion symptoms.

One criticism of this definition of hysteria is that everyone experiences many symptoms characteristic of the syndrome. It is true that most people have experienced headaches, fatigue, anorexia, nausea, diarrhea, vomiting, nervousness, and varied pains. But when responding to a physician, few report such symptoms. Most people interpret the physician's questions to mean significant symptoms; they report only symptoms that are recent, recurrent, or otherwise troublesome. Furthermore, the physician evaluates the patient's responses, ignoring symptoms that are not recent, recurrent, or disabling. He or she does pay attention to symptoms that led the patient to consult a physician, take medicines, or alter usual routines. In addition, there are some symptoms that will be considered significant regardless of qualifying features, such as blindness and paralysis. These criteria—ones that physicians ordinarily use to evaluate symptoms—are the same as those applied in the studies of hysteria cited. By these criteria, very few people report enough symptoms, otherwise unexplained medically, to warrant the diagnosis of hysteria.

Historical Background

The concept of hysteria is at least four thousand years old, and it probably originated in Egypt. The name "hysteria" has been in use since the time of Hippocrates.

The original Egyptian approach to hysteria was probably the most fanciful. Believing that physical displacement of the uterus caused the varied symptoms, physicians treated the patient by trying to attract the "wandering uterus" back to its proper site. Sweet-smelling substances were placed in the region of the vagina to attract the errant organ; unpleasant materials were ingested or inhaled to drive it away from the upper body (58).

Although Egyptian and Greek physicians applied the diagnosis whenever they believed that unusual symptoms were caused by a displaced uterus, they did not provide diagnostic criteria. This state of affairs has persisted to the present and, although speculations about pathogenesis have changed through the centuries, few authors of any century, including our own, have provided diagnostic criteria of any kind (52).

Witchcraft, demonology, and sorcery were associated with hysteria in the Middle Ages (58). Mysterious symptoms, spells, and odd behavior were frequently considered manifestations of supernatural, evil influences. Hysterical patients were sometimes perceived as either the active evil spirit (witch, sorceress, or demon) or as the passive victim of such an evil being. Since the Middle Ages there have been speculations of many kinds about the cause of hysteria. Such speculations have included ideas about neurological weakness, neurological degeneration, the effects of various toxins, and disturbances of what Mesmer called animal magnetism.

Hysteria became Freud's central concern during the early years of psychoanalysis (5). That interest had developed while Freud was working in Paris with Charcot, who was treating hysteria with hypnosis. The psychoanalytic concept of conversion as an ego defense mechanism, referring to unconscious conversion of "psychic energy" into physical symptoms, ultimately resulted in the identification of conversion symptoms with hysteria. Many psychoanalysts consider hysteria a simulation of illness designed to work out unconscious conflicts, partially through attention getting and "secondary gain," a term that refers to possible advantages of illness such as sympathy and support, including financial support, from relatives and friends and being excused from various duties (65).

This view leads easily to the attitude, widely held by physicians though seldom stated openly, that hysteria is a term of opprobrium to be applied whenever the patient's complaints are not explained or when the demands appear excessive. The most fully developed version of this view is the concept of the hysterical personality. Here, the emphasis is on immature, histrionic, manipulative, seductive, and attention-getting behavior regardless of the presenting complaint or present illness. In the absence of specific diagnostic criteria and systematic studies, however, use of the term has been inconsistent and confusing. Many clinicians believe that conversion symptoms, hysteria, and hysterical personality are different: all three may frequently be present in the same individual (33), but any combination is possible and patients may present any one without the other two (10, 11).

The syndromatic approach to the diagnosis of hysteria began in 1859 when the French physician Briquet published his monograph, *Traité clinique et thérapeutique à l'hystérie* (6, 37, 38). Similar descriptions of the syndrome were provided in 1909 by the English physician Savill (50) and in 1951 by the American psychiatrists Purtell, Robins, and Cohen (44). In the past decade a series of studies have refined and clarified the concept of hysteria as a syndrome (26, 28, 29, 43, 63, 64), and in a systematic study of women diagnosed as "having a hysterical personality or as being hysterics," Blinder (3) confirmed the clinical description and familial characteristics described below.

As already noted, the DSM-III-R criteria for somatization disorder and the original criteria for Briquet's syndrome (28) identify somewhat different populations (14, 20). The full significance of these differences is not yet clear. At the same time, a novel statistical approach based on the clustering of certain symptoms in the same individuals, derived from a large-scale epidemiologic study of the prevalence of psychiatric disorders, provided strong support for the "natural occurrence of a polysymptomatic condition resembling somatization disorder" (57).

Epidemiology

A study (61) indicated the prevalence of Briquet's syndrome in an urban community to be 0.4 percent. Assuming that nearly all of the cases occurred in women, and correcting for the sex distribution of the sample, the prevalence of Briquet's syndrome in urban women is just under 1 percent according to this report. Studies of hospitalized, postpartum women whose pregnancies and deliveries were without complication indicate that the prevalence of hysteria is between 1 and 2 percent of the female population (64).

A history of conversion symptoms, on the other hand, is commonly found when systematic interviewing is done among hospitalized, normal, postpartum women (22), hospitalized medically ill women (63), and male and female psychiatric outpatients (31). Table 4.2 indicates that about one-quarter of such patients give a history of conversion symptoms.

All authors, except those describing military experiences, report that the great majority of patients with hysteria are women (19, 32, 34, 46, 56). Nearly all men with symptoms resembling hysteria in women have had histories of associated compensation factors such as litigation following injuries, consideration for veterans' and other pensions, disability payments, or serious legal difficulties (49).

Table 4.2 Prevalence of a history of conversion symptoms

Sample	Percentage
100 normal postpartum women (22)	27
50 medically ill women (63)	30
500 men and women psychiatry clinic patients (31)	24

There is a generally held view that hysteria and conversion symptoms are more common among less sophisticated, more primitive people. One group of investigators found that higher education (one or more years of college) was significantly less common in patients with conversion symptoms—with or without hysteria—than in patients without conversion symptoms. This was also true for patients with hysteria compared to other psychiatric patients as a group (31). Several investigators have reported that "hysterical neurosis, conversion type" is much more frequent in nonwhites than in whites and is highest in the lowest socioeconomic classes (56, 57).

Of much interest is the work of J. C. Carothers, a British psychiatrist whose anthropological studies led to a theory that "psychopathy and hysteria" may be manifestations of the persistence in modern society of preliterate "magical" modes of thinking. On the basis of this theory, Carothers concluded that sociopathy should predominate in men, hysteria in women (see below) (9).

Indeed, although a history of conversion symptoms may be elicited from patients with any psychiatric disorder, two conditions are most often associated with such a history: hysteria and sociopathy.

A report indicates that women with Briquet's syndrome are significantly more likely to show an earlier birth order than are women with other psychiatric disorders (41).

In summary, a history of conversion symptoms can be found among psychiatric patients of either sex suffering from any psychiatric disorder. Such a history is most likely to be associated with hysteria in women or sociopathy in men. Hysteria is much less common than conversion symptoms; it is seen infrequently among men; and, like conversion symptoms, it is infrequently associated with higher education.

Clinical Picture

When first seen by a psychiatrist, the typical patient with hysteria is a married woman in her thirties. Her history is often delivered in a dramatic, complicated fashion. She usually presents with multiple vague complaints to her general physician, and the straightforward history of a present illness is difficult to elicit. Frequently, the physician has trouble deciding when the present illness began or even why the patient came. Table 4.3 presents the most common symptoms reported by hysterics. Not only do hysterics report large numbers of these and other symptoms, but they report symptoms distributed widely throughout all or nearly all organ systems. It is this range of symptoms in addition to their number that defines hysteria.

The dramatic, colorful, exaggerated description of symptoms in hysteria is best conveyed by quoting patients, though it should be noted that not all patients show this trait to the same degree. The following quotations are from Purtell et al. (44):

Vomiting: "I vomit every ten minutes. Sometimes it lasts for two to three weeks at a time. I can't even take liquids. I even vomit water. I can't stand the smell of food."

Food intolerance: "I can't eat pastries. Always pay for it. I can't eat steak now. I throw up whole milk. I always throw up the skins of tomatoes. Pudding made with canned milk makes me sick. I have to use fresh milk."

A trance: "I passed out on the bathroom floor during my period and was still on the floor when they found me the next morning."

Weight change: "I can lose weight just walking down the street. I can hold my breath and lose weight. I was down to 65 pounds at one time."

Dysmenorrhea: "I can't work. Every month I am in bed for several days. I have had to have morphine hypos. There is a throbbing pain in the legs as if the blood doesn't circulate. Can't go to the bathroom as I faint." "It's murder! I want to die. It affects my nervous system."

Sexual indifference: "Have never been interested." "It's not a normal thing to me. Disgusting!" "My husband has never bothered me." "It's just a part of my married life. I have to do it." "I was only disappointed. Never really enjoyed it, but I had to please my husband." "I have no feelings. It's just a duty."

Table 4.3 The frequency of symptoms in hysteria

Symptom	%	Symptom	%	Symptom	%
Dyspnea	72	Weight loss	28	Joint pain	84
Palpitation	60	Sudden fluctuations in weight	16	Extremity pain	84
Chest pain	72	Anorexia	60	Burning pains in rectum, vagina, mouth	28
Dizziness	84	Nausea	80	Other bodily pain	36
Headache	80	Vomiting	32	Depressed feelings	64
Anxiety attacks	64	Abdominal pain	80	Phobias	48
Fatigue	84	Abdominal bloating	68	Vomiting all nine months of pregnancy	20
Blindness	20	Food intolerances	48	Nervous	92
Paralysis	12	Diarrhea	20	Had to quit working because felt bad	44
Anesthesia	32	Constipation	64	Trouble doing anything because felt bad	72
Aphonia	44	Dysuria	44	Cried a lot	60
Lump in throat	28	Urinary retention	8	Felt life was hopeless	28
Fits or convulsions	20	Dysmenorrhea (premarital only)	4	Always sickly (most of life)	40
Faints	56	Dysmenorrhea (prepregnancy only)	8	Thought of dying	48
Unconsciousness	16	Dysmenorrhea (other)	48	Wanted to die	36
Amnesia	8	Menstrual irregularity	48	Thought of suicide	28
Visual blurring	64	Excessive menstrual bleeding	48	Atempted suicide	12
Visual hallucination	12	Sexual indifference	44		
Deafness	4	Frigidity (absence of orgasm)	24		
Olfactory hallucination	16	Dyspareunia	52		
Weakness	84	Back pain	88		

From M. Perley and S. B. Guze (43).

Dyspareunia: "Every time I had intercourse I swelled up on one side. It's sore and burns, and afterwards is very painful." "I hate it. I have a severe pain on the right side and have to go to bed for a day."

Presenting complaints: "I am sore all over. I can't explain it. I have been sick all my life. Now I am alone since my husband died, and the doctor said I must come for help. It has taken $10,000 to keep me alive. This is my 76th hospitalization." "I have been taking care of my invalid mother and I get very little rest or sleep." "My father came here for a checkup on his diabetes and insisted that I come along. I had a nervous breakdown in 1943 and have never gotten around to being really well."

The gynecologist, the neurologist, and the psychiatrist are likely to see hysterics with more focused complaints: the gynecologist because of menstrual pain, irregularity, lapses, or because of dyspareunia; the neurologist because of headaches, "spells," or other conversion symptoms; the psychiatrist because of suicide attempts, depression, or marital discord. But even these specialists find it difficult to obtain straightforward histories from patients with hysteria.

Among the characteristic recurrent or chronic symptoms in hysteria, pains are very prominent: headaches, chest pain, abdominal pain, back pain, joint pain. Abdominal, pelvic, and back pain in association with menstrual or sexual difficulties account for frequent gynecologic surgery: dilatation and curettage, uterine suspension, and salpingo-oopherectomy. Abdominal pain, back pain, dysuria, and dyspareunia account for frequent catheterization and cystoscopy. Abdominal pain, indigestion, bowel difficulties, and vomiting are associated with frequent gastrointestinal X-ray examinations and rectal and gallbladder surgery.

Repeated hospitalization and surgery are characteristic of hysteria (15). Because of the dramatic and persistent symptoms, patients are hospitalized for observation, tests, X-rays, and treatment of a wide variety of medical and surgical conditions that may be mimicked by hysteria. Figures 4.1 and 4.2 illustrate the markedly increased risk of surgery in hysteria.

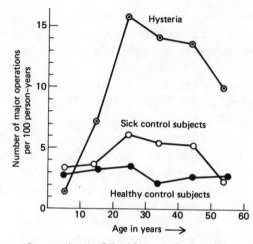

FIGURE 4.1   Comparison of incidence of major operations in hysteria patients with that in medically ill and healthy control subjects. From Cohen et al., (15).

Nervousness and anxiety symptoms (palpitation, dyspnea, chest pain, dizziness, fatigue, tremulousness) are frequent. When chest pain is prominent, it often leads to a diagnosis of heart disease, too often supported only by nonspecific deviations in the electrocardiogram (EKG). The same nervousness and anxiety symptoms, especially when associated with globus (lump in throat) or weight loss, frequently lead to thyroid studies and the diagnosis of thyroid abnormalities. Before the advent of modern laboratory methods for evaluating thyroid function, thyroidectomy was performed for suspected hyperthyroidism, particularly in older patients. Moodiness, irritability, depression, suicidal ideation, and suicide attempts are common and lead to psychiatric hospitalization (17). Hysteria accounts for a significant minority of suicide attempts (51) but rarely leads to suicide (47).

Menstrual symptoms, sexual indifference, and frigidity are so characteristic that a diagnosis of hysteria should be made with care if the menstrual and sexual histories are normal. Marital discord related to sexual indifference and frigidity frequently leads to separation and divorce (26).

50 Hysteria patients   50 Healthy controls

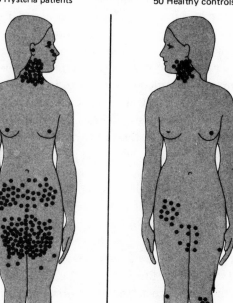

• = 1 operation

FIGURE 4.2 Comparison of number and location of major surgical procedures in fifty hysteria patients and fifty healthy control subjects. By weight, it can be calculated that the mass of organs removed in hysteria patients is more than three times that in control subjects. From Cohen et al., (15).

The question of malingering frequently arises in discussions about hysteria. Although often suspected, malingering is difficult to prove. Nevertheless, malingering and factitious lesions are striking features in some cases of hysteria (45). Factitious fever produced by heating thermometers with matches or friction, skin lesions produced by self-injections, and hemoptysis and hematuria produced by pricking fingers and adding blood to sputum or urine may take months or years to recognize. An occasional patient will confide that a given symptom or sign was produced artificially in the past but will insist that it is now real.

Multiple personality is another manifestation of hysteria, though not all cases of multiple personality appear to be associated with the full syndrome of hysteria. Though considered a rare phenomenon, multiple personality may be increasing in frequency, perhaps as a result of widespread attention in the news media (4, 24).

Another characteristic feature of hysteria is the tendency of many patients to give inconsistent histories (25), so that symptoms that lead to hospitalization on one occasion are denied on another. Though this inconsistency has not been studied systematically, it may be related to the patient's perception of the physician's response to her and her desire to influence his judgment about her illness.

The association between hysteria and sociopathy is of great interest. The familial aspect of this association will be discussed below (see Family Studies), but there is a clinical aspect as well. Many delinquent or antisocial adolescent girls develop hysteria as adults (48); many hysterics give a history of delinquent or antisocial behavior earlier in life (30, 36); the medical histories of delinquent children indicate an increased prevalence of medical contacts (35); and many convicted women felons present with a mixed picture of sociopathy and hysteria (12, 13). Thus school delinquency, repeated fighting, running away from home, a poor work record, a poor marital history, sexual promiscuity, heavy drinking, and police trouble are events found in the histories of many, though not all, hysterics (55). It may be pertinent that Eysenck has concluded from his studies of personality that both hysterics and psychopaths are "extroverted neurotics," a possible reason for associating the two conditions (21).

Natural History

As already noted, it is not easy to determine when the illness began. The vague and often inconsistent history makes the chronology of symptoms difficult to establish. The patient may insist that she has always been "sick" and may describe early difficulties that are hard to evaluate. Hysterics may also suffer from other illnesses, and, therefore, descriptions of childhood or adolescent disorders that allegedly were diagnosed as rheumatic fever, appendicitis, poliomyelitis, or typhoid fever can be most difficult to judge.

The symptoms fluctuate in severity, but the characteristic features persist: recurrent pains, conversion symptoms, nervousness and depression, sexual and marital discord, repeated hospitalization, and repeated surgery (26).

Although systematic follow-up until death of a series of patients has not yet been reported, there is no indication that more than a minority of patients experience marked improvement or permanent remission. Twenty to thirty years of symptoms are typical, and patients who have had the illness for over forty years are not unusual. There is no evidence of excess mortality (18).

On the other hand, the long-term course of the patient with conversion symptoms is determined by the underlying illness rather than by the nature of the symptoms themselves. Because conversion symptoms may be seen in a wide variety of medical, neurologic, and psychiatric disorders (23, 32, 60), patients with these symptoms have a variable prognosis and course.

Complications

The most frequent and important complications of hysteria are repeated surgical operations, drug dependence, marital separation or divorce, and suicide attempts. The first two are presumably preventable if physicians learn to recognize the disorder and manage patients properly. Knowing that hysteria is an alternate explanation for various pains and other symptoms, surgery can be withheld or postponed in cases where objective indications are equivocal or missing. Habit-forming or addicting drugs should be avoided for recurrent or persistent pain.

Whether the physician can modify marital discord through psychotherapy is uncertain, but recognizing the frequency of separation and divorce, the physician can certainly direct his or her attention to these problems. The same may be said about suicide attempts, but there the physician can approach patients with hysteria knowing that the risk of suicide is low.

Family Studies

Hysteria runs in families (1, 16, 62). It was found in one study that about 20 percent of first-degree (siblings, parents, children) female relatives of index cases were hysterics, a tenfold increase over the prevalence in the general population of women, lending strong support to the validity of the concept of hysteria as an illness. A more recent study, using "blind" examiners, showed lower numbers and only a three-and-a-halffold increase in first-degree female relatives of index cases compared to similar relatives of index women with a range of other psychiatric disorders (25), but it again confirmed the familial aggregation of cases.

Family studies also indicate that a significant association exists between hysteria and sociopathy. First-degree male relatives of hysterics show an increased prevalence of sociopathy and alcoholism (1, 25, 62). The first-degree female relatives of convicted male felons show an increased prevalence of hysteria (27). These findings, in conjunction with the observation that many female sociopaths present with, or develop, the full syndrome of hysteria (12), have suggested that at least some cases of hysteria and sociopathy share a common etiology. The widely recognized observation that hysteria is predominantly a disorder of women, whereas sociopathy is predominantly a disorder of men raises the interesting possibility that, depending on the sex of the individual, the same etiologic and pathogenetic factors may lead to different, although sometimes overlapping, clinical pictures.

Two studies (7, 8, 59) of adopted children whose biological parents showed antisocial behavior revealed greater than normal frequency in female offspring of hysteria or other multiple unexplained somatic complaints, thus supporting the hypothesis that hysteria and sociopathy are related disorders. One of these studies was carried out in the United States; the other in Sweden.

Differential Diagnosis

Three psychiatric conditions must, at times, be considered in the differential diagnosis of hysteria: panic disorder, major depressive disorder, and schizophrenia.

As already noted, the characteristic symptoms of hysteria include many that are also seen in panic disorder and depression. If the patient is a young woman with menstrual and sexual difficulties who presents with a full range of anxiety or depressive symptoms, she may nearly meet the research criteria for hysteria. Age of onset, details of the symptom picture, course, and mental status will usually help clarify the diagnostic problem.

A number of patients with schizophrenia also meet the diagnostic criteria for hysteria. Also, an occasional patient with apparent hysteria who does not show definite evidence of schizophrenia when first seen will, in time, develop the typical clinical picture of schizophrenia. All the patients seen thus far with a combined hysteria-schizophrenia picture have eventually suffered from prominent and usually systematized delusions (64).

Unlike the situation in which the same patients may meet diagnostic criteria for sociopathy and hysteria, a situation in which family studies suggest that the two disorders are associated, the overlap between hysteria and schizophrenia is not accompanied by evidence of a familial association.

Clinical Management

The diagnosis of hysteria provides two advantages. For patients with the full diagnostic picture, the physician can predict in over 90 percent of cases that the characteristic symptoms will continue through time and that other illnesses, which in retrospect could account for the original clinical picture, will not become evident. For patients who do not present with the full diagnostic picture, especially for those who present with conversion symptoms and little else, the physician knows that in a substantial number of cases other disorders will become evident that do, in retrospect, account for the original clinical picture (43, 60). Many, perhaps most, of the patients who resemble hysterics, but fail to meet the diagnostic criteria, will remain undiagnosed at follow-up. Enough of them, however, will turn out to have other serious medical, neurological, or psychiatric disorders to justify the kind of diagnostic open-mindedness that can lead to earlier recognition of the other illnesses.

Patients with hysteria are difficult to treat (2, 42). The typical clinical picture of recurrent, multiple, vague symptoms combined with doctor-shopping and frequent requests for time and attention can frustrate and anger a physician. Few hysterics referred to a psychiatrist persist with psychiatric treatment (26). Thus the burden of caring for these patients continues to rest with other physicians. A controlled, randomized study indicates, however, that psychiatric consultation may be helpful in reducing the extent and cost of medical care (54).

The physician's major goal should be to avoid the complications described earlier. Success will depend on winning the patient's confidence without allowing the patient's symptomatic behavior to exhaust the physician's sympathy. The patient and family can be told that the patient tends to experience symptoms that suggest other disorders, but that they are not medically serious. On this basis, the physician can approach each new complaint circumspectly and conservatively, especially with regard to elaborate or expensive diagnostic studies. As he grows familiar with the patient, the physician will become increasingly confident in his clinical judgments. Remembering that hysteria can "explain" many puzzling symptoms, the physician, always looking for objective evidence of other disorders, can avoid unnecessary hospitalization and surgery.

The physician should strive to substitute discussions of the patient's life problems, personality, and concerns for a quick-triggered response of additional tests and X-rays or yet another drug. At the same time, he must recognize the limitations resulting from the lack of knowledge concerning the causes of the disorder.

References

1. Arkonac, O., and Guze, S. B. A family study of hysteria. N. Engl. J. Med. 268:239–242, 1963.
2. Bird, J. The behavioural treatment of hysteria. Brit. J. Psychiat. 134: 129–137, 1979.
3. Blinder, M. G. The hysterical personality. Psychiatry 29:227–235, 1966.
4. Boor, M. The multiple personality epidemic. Additional cases and inferences regarding diagnosis, etiology, dynamics, and treatment. J. Nerv. Ment. Dis. 170:302–304, 1982.
5. Breuer, J., and Freud, S. Studies in Hysteria (trans. A. A. Brill). New York: Nerv. Ment. Dis. Monogr., 1936.
6. Briquet, P. Traité clinique et thérapeutique à l'hysténe. Paris: J.-B. Ballière & Fils, 1859.
7. Cadoret, R. J. Psychopathology in adopted-away offspring of biologic parents with antisocial behavior. Arch. Gen. Psychiat. 35:176–184, 1978.
8. Cadoret, R. J., Cunningham, L., Loftus, R., and Edwards, J. Studies of adoptees from psychiatrically disturbed biological parents. III. Medical symptoms and illness in childhood and adolescence. Am. J. Psychiat. 133: 1316–1318, 1976.
9. Carothers, J. C. Hysteria, psychopathy and the magic word. Mankind Q. 16:93–103, 1975.
10. Chodoff, P. The diagnosis of hysteria: an overview. Am. J. Psychiat. 131:1073–1078, 1974.
11. Chodoff, P., and Lyons, H. Hysteria: the hysterical personality and hysterical conversion. Am. J. Psychiat. 114:734–740, 1958.
12. Cloninger, C. R., and Guze, S. B. Psychiatric illness and female criminality: the role of sociopathy and hysteria in the antisocial woman. Am. J. Psychiat. 127:303–311, 1970.
13. Cloninger, C. R., and Guze, S. B. Female criminals: their personal, familial, and social backgrounds. Arch. Gen. Psychiat. 23:554–558, 1970.
14. Cloninger, C. R., Martin, R. L., Guze, S. B., and Clayton, P. J. A prospective follow-up and family study of somatization in men and women. Am. J. Psychiat. 143:873–878, 1986.
15. Cohen, M. E., Robins, E., Purtell, J. J., Altmann, M. W., and Reid, D. E. Excessive surgery in hysteria. JAMA 151:977–986, 1953.
16. Coryell, W. A blind family history study of Briquet's syndrome. Further validation of the diagnosis. Arch. Gen. Psychiat. 37:1266–1269, 1980.
17. Coryell, W., and Norten, S. G. Briquet's syndrome (Somatization Disorder) and primary depression: comparison of background and outcome. Compr. Psychiat. 22:249–256, 1981.
18. Coryell, W. Diagnosis-specific mortality. Primary unipolar depression and Briquet's syndrome (somatization disorder). Arch. Gen. Psychiat. 38:939–942, 1981.
19. De Figueiredo, J. M., Baiardi, J. J., and Long, D. M. Briquet syndrome

in a man with chronic intractable pain. Johns Hopkins Med. J. 147:102–106, 1980.

20. DeSouza, C., and Othmer, E. Somatization disorder and Briquet's syndrome. Arch. Gen. Psychiat. 41:334–336, 1984.

21. Eysenck, H. The Dynamics of Anxiety and Hysteria. New York: Praeger, 1957.

22. Farley, J., Woodruff, R. A., Jr., and Guze, S. B. The prevalence of hysteria and conversion symptoms. Brit. J. Psychiat. 114:1121–1125, 1968.

23. Gatfield, P. D., and Guze, S. B. Prognosis and differential diagnosis of conversion reactions. Dis. Nerv. Syst. 23:623–631, 1962.

24. Greaves, G. B. Multiple personality. 165 years after Mary Reynolds. J. Nerv. Ment. Dis. 168:577–596, 1980.

25. Guze, S. B., Cloninger, C. R., Martin, R. L., and Clayton, P. J. A follow-up and family study of Briquet's syndrome. Brit. J. Psychiat. 149:17–23, 1986.

26. Guze, S. B., and Perley, M. J. Observations on the natural history of hysteria. Am. J. Psychiat. 119:960–965, 1963.

27. Guze, S. B., Wolfgram, E. D., McKinney, J. K., and Cantwell, D. P. Psychiatric illness in the families of convicted criminals: a study of 519 first-degree relatives. Dis. Nerv. Syst. 28:651–659, 1967.

28. Guze, S. B. The diagnosis of hysteria: What are we trying to do? Am. J. Psychiat. 124:491–498, 1967.

29. Guze, S. B. The role of follow-up studies: their contribution to diagnostic classification as applied to hysteria. Semin. Psychiat. 2:392–402, 1970.

30. Guze, S. B., Woodruff, R. A., Jr., and Clayton, P. J. Hysteria and antisocial behavior: further evidence of an association. Am. J. Psychiat. 127:957–960, 1971.

31. Guze, S. B., Woodruff, R. A., and Clayton, P. J. A study of conversion symptoms in psychiatric outpatients. Am. J. Psychiat. 128:643–646, 1971.

32. Hafeiz, H. B. Hysterical conversion: a prognostic study. Brit. J. Psychiat. 136:548–551, 1980.

33. Kimble, R., Williams, J. G., and Agras, S. A comparison of two methods of diagnosing hysteria. Am. J. Psychiat. 132:1197–1199, 1975.

34. Kroll, P., Chamberlain, K. R., and Halpern, J. The diagnosis of Briquet's syndrome in a male population. The Veterans Administration revisited. J. Nerv. Ment. Dis. 167:171–174, 1979.

35. Lewis, D. O., and Shanok, S. S. Medical histories of delinquent and nondelinquent children: an epidemiologic study. Am. J. Psychiat. 134:1020–1025, 1977.

36. Lilienfeld, S. O., Van Valkenburg, C., Larntz, K., and Akiskal, H. S. The relationship of histrionic personality disorder to antisocial personality and somatization disorders. Am. J. Psychiat. 143:718–722, 1986.

37. Mai, F. M., and Merskey, H. Briquet's Treatise on Hysteria. A synopsis and commentary. Arch. Gen. Psychiat. 37:1401–1405, 1980.

38. Mai, F. M., and Merskey, H. Historical review. Briquet's concept of hysteria: an historical perspective. Can. J. Psychiat. 26:57–63, 1981.

39. Marsden, C. D. Hysteria—a neurologist's view. Psychol. Med. 16:277–288, 1986.
40. Mohr, P. D., and Bond, M. J. A chronic epidemic of hysterial blackouts in a comprehensive school. Brit. Med. J. 284:961–962, 1982.
41. Morrison, J. R. Early birth order in Briquet's syndrome. Am. J. Psychiat. 140:1596–1597, 1983.
42. Murphy, G. E. The clinical management of hysteria. JAMA 247:2559–2564, 1982.
43. Perley, M. J., and Guze, S. B. Hysteria—the stability and usefulness of clinical criteria. N. Engl. J. Med. 266:421–426, 1962.
44. Purtell, J. J., Robins, E., and Cohen, M. E. Observations on clinical aspects of hysteria. JAMA 146:902–909, 1951.
45. Ries, R. K. Single case study. DSM-III differential diagnosis of Munchausen's syndrome. J. Nerv. Ment. Dis. 168:629–632, 1980.
46. Robins, E., Purtell, J. J., and Cohen, M. E. "Hysteria" in men. N. Engl. J. Med. 246:677–685, 1952.
47. Robins, E., Murphy, G., Wilkinson, R., Gassner, S., and Kayes, J. Some clinical considerations in the prevention of suicide based on a study of 134 successful suicides. Am. J. Public Health 49:888–899, 1959.
48. Robins, L. Deviant Children Grown Up. Baltimore: Williams & Wilkins, 1966.
49. Rounsaville, B. J., Harding, P. S., and Weissman, M. M. Single case study. Briquet's syndrome in a man. J. Nerv. Ment. Dis. 167:364–367, 1979.
50. Savill, T. D. Lectures on Hysteria and Allied Vasomotor Conditions. London: H. J. Glaisher, 1909.
51. Schmidt, E. H., O'Neal, P., and Robins, E. Evaluation of suicide attempts as guide to therapy. Clinical and follow-up study of 109 patients. JAMA 155:549–557, 1954.
52. Slater, E. Diagnosis of "hysteria." Brit. Med. J. 1:1395–1399, 1965.
53. Small, G. W., and Nicholi, A. M., Jr. Mass hysteria among schoolchildren. Early loss as a predisposing factor. Arch. Gen. Psychiat. 39:721–724, 1982.
54. Smith, G. R., Jr., Monson, R. A., and Ray, D. C. Psychiatric consultation in somatization disorder. N. Engl. J. Med. 314:1407–1413, 1986.
55. Spalt, L. Hysteria and antisocial personality. A single disorder? J. Nerv. Ment. Dis. 168:456–464, 1980.
56. Stefansson, J. G., Messina, J. A., and Meyerowitz, S. Hysterical neurosis, conversion type: clinical and epidemiological considerations. Acta Psychiat. Scand. 53:119–138, 1976.
57. Swartz, M., Blazer, D., Woodbury, M., George, L., and Landerman, R. Somatization disorder in a U.S. southern community: use of a new procedure for analysis of medical classification. Psychol. Med. 16:595–609, 1986.
58. Vieth, I. Hysteria. The History of a Disease. Chicago: Univ. of Chicago Press, 1965.

59. von Knorring, A.-L. Adoption studies on psychiatric illness. Umea University Medical Dissertations, Series no. 101. Umea: Sweden, 1983.
60. Watson, C. G., and Buranen, C. The frequency and identification of false positive conversion reactions. J. Nerv. Ment. Dis. 167:243–247, 1979.
61. Weissman, M. M., Myers, J. K., and Harding, P. S. Psychiatric disorders in a U.S. urban community: 1975–76. Am. J. Psychiat. 135:459–462, 1978.
62. Woemer, P. I., and Guze, S. B. A family and marital study of hysteria. Brit. J. Psychiat. 114:161–168, 1968.
63. Woodruff, R. A., Jr. Hysteria: an evaluation of objective diagnostic criteria by the study of women with chronic medical illnesses. Brit. J. Psychiat. 114:1115–1119, 1967.
64. Woodruff, R. A., Jr., Clayton, P. J., and Guze, S. B. Hysteria: studies of diagnosis, outcome, and prevalence. JAMA 215:425–428, 1971.
65. Ziegler, F. J., Imboden, J. B., and Meyer, E. Contemporary conversion reactions: clinical study. Am. J. Psychiat. 116:901–910, 1960.

5. Obsessive Compulsive Disorder

Definition

Obsessions are persistent distressing thoughts or impulses experienced as unwanted and senseless but irresistible. *Compulsions* are acts resulting from obsessions. Obsessive compulsive disorder is a chronic illness dominated by obsessions and compulsions occurring in the absence of another psychiatric disorder. Obsessional neurosis is a synonym.

Historical Background

The term "obsessional neurosis" apparently originated with Karl Westphal (1833–90), a German neurologist who wrote about obsessional conditions and phobias. Kraepelin described obsessional neurosis (*Zwangsneurose*) in his textbooks of the early part of the twentieth century, and the same term was adopted by Freud, whose classical description of the clinical picture, published in 1917, follows (13):

> The obsessional neurosis takes this form: the patient's mind is occupied with thoughts that do not really interest him, he feels impulses which seem alien to him, and he is impelled to perform actions which not only afford him no pleasure but from which he is powerless to desist. The thoughts (obsessions) may be meaningless in themselves or only of no interest to the patient; they are often only absolutely silly.

Recognition of obsessional traits, however, far antedated psychiatrists' description of the syndrome. In the early seventeenth century Richard Flecknoe, in discussing "enigmaticall characters," described one such character as an "irresolute Person" (20):

He hovers in his choice, like an empty Ballance with no waight of Judgement to incline him to either scale . . . everything he thinks on, is matter of deliberation . . . and he does nothing readily, but what he thinks not on . . . when he begins to deliberate, never makes an end. . . . Has some dull demon cryes, do not, do not still, when hee's on point of doing anything. . . . He plays at shall I, shall I? so long, till opportunity be past . . . and then repents at leisure.

In the seventeenth century obsessions were often referred to as "scruples," defined by Jeremy A. Taylor (52) as

a great trouble of mind preceding from a little motive, and a great indisposition, by which the conscience though sufficiently determined by proper arguments, dares not proceed to action, or if it do, it cannot rest . . . some persons dare not eat for fear of gluttony, they fear that they shall sleep too much, and that keeps them waking, and troubles their heads more and then their scruples increase. When they are married they are afraid to do their duty, for fear it be secretly an indulgence to the flesh, and to be suspected of carnality, and yet they dare not omit it, for fear they should be unjust, and yet their fear that the very fearing it to be unclean should be a sin, and suspect that if they do not fear so it is too great a sign they adhere to Nature more than to the Spirit. . . . Scruple is a little stone in the foot, if you set it upon the ground it hurts you; if you hold it up you cannot go forward; it is a trouble where the trouble is over, a doubt when doubts are resolved . . . very often it hath no reason at all for its inducement.

The content of obsessions, today as in years past, is often religious. In his monograph Of Religious Melancholy (40), published in 1692, John Moore described "naughty and sometime blasphemous thoughts which start in the mind while they are exercised in the worship of God despite all their endeavors to stiffle and suppress them." In fact, he wrote, "the more they struggle with them, the more they increase." Bishop Moore had difficulty understanding this state as the sufferers were "mostly good People," whereas "bad men . . . rarely know anything of these kinds of Thoughts." Like others, he argued that this was good reason "to judge them to be Distempers of the body rather than Faults of the Mind." He was particularly concerned about the phobic avoidances that could result from such obsessions and wrote, "I exhort you not to quit your Employment . . . for no business at all is as bad for you as too much: and there is always more Melancholy to be found in a Cloyster than in the Marketplace."

The obsessional fear of suffering from syphilis was also well recognized as a manifestation of mental illness by the clergy and later by psychiatrists. In the past, "superstition" has often been used to mean what we would consider an obsession. Samuel Johnson apparently had this in mind when he wrote that "the superstitious are often melancholic, and the melancholic almost always superstitious." As will be noted later, obsessions are common in depressive states and vice versa.

Epidemiology

Obsessive compulsive disorder has been regarded as rare, but a mid-1980s study suggests that the disorder may affect 2 percent of the general population, making it one of the more common forms of serious psychopathology (42, 44). If true, only a small proportion of obsessionals visit psychiatrists. Less than 5 percent of psychiatric inpatients and outpatients receive the diagnosis. According to most studies, obsessional disorder occurs about equally in both sexes.

When compared with all other psychiatric patients, obsessionals reportedly differ in the following ways: (a) they belong to higher social classes, (b) they make higher scores on intelligence tests, (c) they have a higher educational level (21, 34, 46). Since social class, intelligence, and educational level are interrelated, predictably if one is elevated, the others will be also. Obsessionals are often in their early or mid thirties when first admitted to the hospital (5).

Clinical Picture

When obsessionals finally seek medical attention (often years after their illness began), it may be because of depression, acute anxiety, exacerbation of obsessions, or social incapacity resulting from any of these conditions (33, 41) (Table 5.1).

Table 5.1 Diagnostic criteria for Obsessive Compulsive Disorder
(DSM-III-R)

A. Either obsessions or compulsions:

Obsessions:

(1) recurrent and persistent ideas, thoughts, impulses, or images
that are experienced, at least initially, as intrusive and senseless, e.g., a parent's having repeated impulses to kill a loved
child, a religious person's having recurrent blasphemous
thoughts

(2) the person attempts to ignore or suppress such thoughts or
impulses or to neutralize them with some other thought or
action

(3) the person recognizes that the obsessions are the product of
his or her own mind, not imposed from without (as in
thought insertion)

(4) if another disorder is present, the content of the obsession
is unrelated to it, e.g., the ideas, thoughts, impulses, or
images are not about food in the presence of an Eating Disorder, about drugs in the presence of a Psychoactive Substance Use Disorder, or guilty thoughts in the presence of a
Major Depression

Compulsions:

(1) repetitive, purposeful, and intentional behaviors that are
performed in response to an obsession, or according to certain rules, or in a stereotyped fashion

(2) the behavior is designed to neutralize or to prevent discomfort or some dreaded event or situation; however, either the
activity is not connected in a realistic way with what it is
designed to neutralize or prevent, or it is clearly excessive

(3) the person recognizes that his or her behavior is excessive or
unreasonable (this may not be true for young children; it
may no longer be true for people whose obsessions have
evolved into overvalued ideas)

B. The obsessions or compulsions cause marked distress, are time-
consuming (take more than an hour a day), or significantly interfere with the person's normal routine, occupational functioning, or usual social activities or relationships with others.

A life event often seems associated with the onset of symptoms: the death of a relative, a sexual conflict, overwork, or pregnancy (21, 34). Just as often, however, no precipitating factors can be identified.

A common presenting complaint is the obsessional fear of injuring oneself or another person, often a child or close relative. Fearful of losing control, the patients may develop avoidances or rituals that lead in turn to social incapacity. Perhaps they will refuse to leave the house or will avoid sharp objects or wash repeatedly.

On reflection they may identify the obsessional idea as illogical, but not always. Sometimes the ideas are not, strictly speaking, illogical (germs do produce disease) and sometimes, even when obviously absurd, the ideas are not viewed as such. What distinguishes an obsession from a delusion is not so much insight (recognizing the idea's absurdity) as the person's struggle against the obsessional experience itself. He strives to free himself from the obsession but cannot and feels increasingly uncomfortable until the idea temporarily runs its course or the obsessional act has been completed (33).

The frequency with which obsessional symptoms appear in obsessional patients has been studied systematically by three groups, and their results indicate that the illness may assume one or more of the following forms (1, 10, 51):

Obsessional ideas. Thoughts that repetitively intrude into consciousness (words, phrases, rhymes) and interfere with the normal train of thought and causing distress to the person. Often the thoughts are obscene, blasphemous, or nonsensical.

Obsessional images. Vividly imagined scenes, often of a violent, sexual, or disgusting nature (images of a child being killed, cars colliding, excrement, parents having sexual intercourse) that repeatedly come to mind.

Obsessional convictions. Notions that are often based on the magical formula of thought equals act: "Thinking ill of my son will cause him to die." Unlike delusions, obsessional beliefs are characterized by ambivalence: The person believes and simultaneously does not believe. As Jaspers expressed it, there is a "constant going on between a consciousness of validity and non-validity. Both push this way and that, but neither can gain the upper hand" (26).

Obsessional rumination. Prolonged, inconclusive thinking about a subject to the exclusion of other interests. The subject is often religion or metaphysics—why and wherefore questions that are as unanswerable as they are endlessly ponderable. Indecisiveness in ordinary matters is very common: "Which necktie should I wear?" Doubt may lead to extremes in caution, both irksome and irresistible. "Did I turn off the gas?" "Lock the door?" "Write the correct address?" The patient checks and rechecks, stopping only when exhausted or on checking a predetermined "magical" number of times. Several studies suggest that obsessional doubts— *manie du doute*—may well be the most prominent feature in obsessional neurosis (1, 10, 48). As with other obsessions, ruminations are resisted. The person tries to turn his or her attention elsewhere, but cannot; often the more he or she tries, the more intrusive and distressing the thoughts become.

Obsessional impulses. Typically relates to self-injury (leaping from a window); injury to others (smothering an infant); or embarrassing behavior (shouting obscenities in church).

Obsessional fears. Often of dirt, disease, contamination; of potential weapons (razors, scissors); of being in specific situations or performing particular acts.

Obsessional rituals (compulsions). Repetitive, stereotyped acts of counting, touching, arranging objects, moving, washing, tasting, looking in specific ways. Compulsions are inseparable from the obsessions from which they arise. A compulsion is an obsession expressed as action. About one-quarter of patients display no compulsions (1).

Counting rituals are especially common. The person feels compelled to count letters or words or the squares in a tile floor or to perform arithmetical operations. Certain numbers, or their multiples, may have special significance (the person "must" lay down her pencil three times or step on every fifth sidewalk crack).

Other rituals concern the performance of excretory functions and such everyday acts as preparing to go to bed. Also common are rituals involving extremes of cleanliness (handwashing compulsions, relentless emptying of ashtrays) and complicated routines assuring orderliness and punctuality. Women apparently have a higher incidence of contamination phobia and of compulsive cleaning behavior than do men (10).

According to a study of obsessionals (51), four kinds of rituals occur most frequently: counting, checking, cleaning, and avoidance rituals. Each occurred in half of the patients. Avoidance rituals were similar to those seen in phobic disorders (see Differential Diagnosis). An example can be seen in a patient who avoided anything colored brown. Her inability to approach brown objects greatly limited her activities.

Other rituals that occur less often consist of slowing, striving for completeness, and extreme meticulousness. With slowing, such simple tasks as buttoning a shirt or tying a shoelace might take up to fifteen minutes. Striving for completeness may be seen in dressing also. Asked why he spends so much time with a single button, the patient might reply that he was trying to prove to himself that he had "buttoned the button properly."

A common form of pathological meticulousness is a concern that objects be arranged in a special way. Pencils, for example, may have to be arranged so that the points are directed away from the patient. Students may spend so much time in arranging pencils, pens, erasers, and so on, that they cannot do their work.

Rituals, ridiculous as they may seem to the patients, are accompanied by a profound dread and apprehension that assure their performance because they alone give relief. "I'll explode if I don't do it," a patient may say. Occasionally patients believe that failure to perform a given ritual will result in harm to themselves or others; often the ritual is as inexplicable to patients as it is to observers.

Obsessional symptoms often are accompanied by a dysphoric mood. The patient may be irritable, tense, depressed. This may lead to an erroneous diagnosis of affective disorder because the mood element at the time of examination may overshadow the obsessional content.

Obsessional symptoms rarely occur singly (30). As with most psychiatric illnesses, obsessional disorder presents a cluster of symptoms that, individually, are variable and inconstant over time but as a group maintain characteristics unique to the illness. Thus a patient may now have one set of obsessional impulses, phobias, and rituals and later another set, but the symptoms remain predominantly obsessional.

Natural History

Most obsessionals develop their illness before age twenty-five (9). Many obsessionals have clear-cut obsessional symptoms by age fifteen and the illness may begin as early as age six. Fewer than 15 percent of obsessionals develop their illness after age thirty-five (17, 21, 30).

Though the mean age of onset of symptoms is roughly twenty in both sexes, the first psychiatric contact is made on the average about seven years later. If hospitalized, most patients are in their thirties by the time they enter a hospital. Rarely are obsessionals admitted to a psychiatric hospital for the first time after age forty (16).

The mode of onset may be acute or insidious (30). The course of obsessional disorder may be unremitting (with or without social incapacity), episodic, or characterized by incomplete remissions that permit normal social functioning.

In one outpatient study, the majority of patients had episodic courses with exacerbations that usually lasted less than a year (41). Most investigators who study patients requiring hospitalization, however, have found the course to be steady, with exacerbations often attributed to fatigue or medical illness and with a tendency for the severity of the illness to wane gradually over many years (21, 34).

Obsessionals with mild symptoms requiring only outpatient therapy appear to have a rather good prognosis; as many as 60 to 80 percent are asymptomatic or improved one to five years after diagnosis (34, 41). Hospitalized cases as a group do less well. One-third or fewer are improved symptomatically on reexamination several years after discharge; two-thirds or more, however, are functioning as well socially as before hospitalization (21, 30). Between 5 and 10 percent of clear-cut obsessionals do have a course marked by progressive social incapacity (45). Favorable prognosis is reported to be associated with three factors: (a) mild or atypical symptoms, including predominance of phobic-ruminative ideas and absence of compulsions (21, 30); (b) short duration of symptoms before treatment is begun (41); (c) good premorbid personality without childhood symptoms or abnormal personality traits (30, 31). The specific content of obsessions is not believed to have prognostic significance (Table 5.2).

Complications

Depression—often indistinguishable symptomatically from depression seen in primary affective disorder—is probably the most common complication of obsessional disorder.

Failure to marry also may be a complication of obsessional disorder, judging by three studies (21, 34, 46) that show a higher rate of celibacy among obsessionals than in the general population.

Despite the frequency with which suicide may figure in obsessional thinking (17, 21), obsessionals rarely commit suicide. In most studies less than 1 percent of the patients committed suicide.

Obsessional patients sometimes fear they will injure someone by an impulsive act. They fear they will lose control and embarrass themselves in some manner. They worry about becoming addicted to drugs prescribed by their physician. These fears are generally unwarranted (9). There is little evidence that obsessional disorder predisposes to homicide, criminal behavior, alcoholism, or drug addiction (21, 30).

Finally, obsessionals may fear they will "lose" their minds, become totally disabled, need chronic hospitalization. None of these

Table 5.2 Follow-up studies of obsessional disorder

	Sample Characteristics			Length of Follow-up (to nearest yr.)	Condition on Follow-up (%)		
Author	Place	Pt. Source*	N†		Asymptomatic	Improved	Unimproved
Balslev-Olesen et al. (5)	Denmark	I,O	52	0-8	6	58	37
Coryell (9)	U.S.	I	36	1-32	22	57	22
Grimshaw (17)	England	O	97	1-14	40	24	35
Hastings (18)	U.S.	I	23	6-12	13	40	47
Ingram (21)	England	I	46	1-11	9	30	61
Kringlen (30)	Norway	I	85	13-20	4	45	45
Langfeldt (31)	Norway	I	27	1-11	26	41	33
Lewis (33)	England	I,O	50	>5	32	34	34
Lo (34)	Hong Kong	I,O	87	1-14	20	36	44
Luff et al. (35)	England	O	49	3	39	27	34
Pollitt (41)	England	I,O	66	0-15	24	48	28
Rennie (43)	U.S.	I	47	20	36	38	26
Rüdin (46)	Germany	I,O	130	2-26	12	26	61

* I = inpatient; O, outpatient.
† Excluded are lobotomized cases and patients dead on follow-up. The sex ratio differs from study to study but approaches unity when all studies are combined.

Source: Goodwin et al. (10)

events is a common complication of obsessional disorder. If schizophrenia is clearly ruled out at the beginning, obsessionals probably become schizophrenic no more often than nonobsessionals. They infrequently become totally disabled, and they usually do not require long-term hospitalization (17, 21, 41).

Family History

Two studies comparing parents of obsessionals with parents of other psychiatric patients found the former group to be significantly more perfectionistic (34), obstinate, pedantic, and parsimonious (30). Studies without controls have yielded similar findings (5, 33). Unfortunately, such attributes as perfectionism and obstinacy are vague and resist quantification. Even in controlled studies, it is unclear whether observer bias has been sufficiently eliminated to render the results meaningful.

Psychiatric illness in the families of obsessionals has been studied by several investigators. Many obsessionals have close relatives with psychiatric illness (5, 19, 45). In one study, about 30 percent of the siblings of obsessionals had been treated for psychiatric symptoms (about one-quarter as inpatients) (33) and in another study, about 10 percent of the first-degree relatives of obsessionals had received psychiatric treatment (45).

Obsessional illness probably occurs with greater frequency in the families of obsessionals than in the families of other patients (or in the general population), but so do other psychiatric illnesses, and the extent of increased familial risk is uncertain. One study comparing the families of obsessionals with those of other psychiatric and medical patients found not only a higher incidence of obsessional disorder in the families of obsessionals (8 percent of parents, 7 percent of siblings) but also a higher incidence of other psychopathology, for example, about 8 percent had manic depressive disease (6). In another study (19), no instances of obsessive compulsive disorder were found in 174 relatives of 10 severely obsessive compulsive patients, although 12 percent of first-degree relatives had been hospitalized for psychiatric illness. Some evidence suggests a link to Tourette's syndrome (28, 32). About half of people with Tourette's syndrome also have obsessive compulsive symptoms along with better-known manifestations of Tourette's syndrome such as tics and bizarre verbalizations.

Studies of twins with obsessional illness indicate that monozygotic twins are more concordant for obsessional illness than dizygotic twins (22). About 80 to 90 percent of monozygotic twins are concordant for obsessional illness versus a concordance rate in dizygotic twins of no more than 50 percent and probably much lower. This does not prove the existence of genetic factors in obsessional illness; all of the twin pairs were raised together and identical twins may emulate each other more than fraternal twins. However, in instances where twins have little contact, environmental explanations seem implausible, as illustrated by the following case histories adapted from a published report (39).

The "W" twins
Jean, a housewife, developed her symptoms at 24 when she acquired a fear of contamination by dirt with severe washing and cleansing rituals. At the time of her referral to this hospital three years later she was washing her hands between 60 and 80 times per day, spending 12 hours a day cleaning and disinfecting her house, and using up to 20 quarts of liquid disinfectant a week. Her daughter, aged 2½ years, was not toilet trained and was restricted to one room of the house in order to avoid spread of contamination, and her husband had been obliged to give up sports for fear that his soiled sports clothes would introduce dirt to their home. On admission to the hospital the skin of her hands was roughened, red, cracked and bleeding.

Jill, the co-twin, a social worker, developed her symptoms at 22 when she and Jean were leading separate lives and had little contact with each other. Her traits of neatness and cleanliness became more pronouncd and she developed rituals concerned with washing up the crockery and cutlery which had to be started immediately when a meal was finished, any delay resulting in severe anxiety; the washing of the dishes and utensils had to proceed in a specific order and failure to comply with the routine, or an attempt by others to relieve her of her task, provoked great discomfort. She attempted to resist carrying out her rituals which she considered silly and unnecessary but the time spent on them increased and interfered with her social life.

Their father, a retired clerk, was fastidiously neat and orderly in his habits but neither he nor other family members had received psychiatric treatment. Their early development and childhood were unremarkable, and at school both did well academically. They attended separate universities. Both were normally outgoing and sociable and both had personality traits of orderliness, determination and conscientiousness. Neither had suffered any previous illness and they only learned of their similar obsessional problems after Jean began treatment.

The "K" twins

Linda, a secretary, began having obsessional thoughts at 25 when she felt envious of a roommate who had formed a close relationship with a man. Aware of her envy and associated guilt, she began to think of imaginary pleasant scenes of herself with her friend, but within a few weeks these thoughts became recurrent, intrusive and difficult to resist and were no longer associated with relief of unpleasant feelings. Obsessional thoughts troubled her to a varying degree for several years, with exacerbations during times of stress or unhappiness. Often she would have to repeat a conversation in her mind and, if interrupted before it was completed, she felt bound to repeat the whole conversation again. Sometimes she was compelled to touch an object to signify completion of an obsessional thought and this sometimes had a magical aspect; for example, if the object was a clock, the numbers indicating the time became her "lucky numbers" for the rest of the day.

Ann, the co-twin, an actress, also developed obsessional thoughts in her mid-twenties. She was not in regular contact with Linda at this time and only years later learned of her similar difficulties. Ann also used "good thoughts" to counter anxiety which was usually caused by parental disapproval of her career. The thoughts soon developed an obsessional quality and were associated with compulsive behaviour. For example, if a "bad thought" occurred to her while speaking on the telephone, after the call was ended she felt compelled to repeat the conversation and create "good thoughts." If she passed a shop window and could not recall its contents she would have to go back and check them and she often spent hours arranging clothing in her wardrobe according to the color or type of garment. Although she considered her obsessions to be irrational, any attempts to resist them gave rise to the fears of impending disaster.

The father was an accountant described as being "compulsively neat." Their mother had older twin sisters who both had a life-long phobia of birds.

Differential Diagnosis

Obsessions occur in children, in healthy adults, and in patients with a variety of psychiatric and medical illnesses.

The rituals and superstitions of children—avoidance of sidewalk cracks, insistence on a given routine, carrying of amulets and charms—may resemble the compulsive acts of obsessional disorder, but with a difference: Children usually do not complain about these acts, which seem natural to them and produce little distress.

Only a small proportion of children manifesting obsessional behavior can be classified by usual standards as having obsessional disorder. It is not known whether children who are exceptionally ritualistic and superstitious have a greater risk of developing obsessional disorder than their less obsessional peers.

Many adult obsessionals give a history of obsessional symptoms in childhood (21, 34, 46), but the commonness of such symptoms and the influence of retrospective distortion make such data hard to interpret. Obsessionals commonly cite phobias and rituals as childhood symptoms (21, 30, 34). They rarely give a history of stealing, truancy, or tantrums in childhood (30, 34).

Obsessional personality is more a description than a diagnosis. No investigator has followed a group of clearly defined obsessional personalities over time to determine their fate; hence the label has no predictive value and is not in this sense a diagnosis. The individual with an obsessional personality is punctual, orderly, scrupulous, meticulous, and dependable. He is also rigid, stubborn, pedantic, and something of a bore. He has trouble making up his mind, but once made up, is single-minded and obstinate (49).

Many individuals with obsessional disorder have obsessional personalities antedating the illness (30, 41, 46). The proportion of obsessional personalities who develop obsessional disorder, depression, or other psychiatric illness is not known.

Phobias are commonly associated with obsessional disorder, and obsessionals often have anxiety symptoms. Obsessional phobias have a compulsive quality and are almost always accompanied by rituals as well as other obsessional phenomena (30). The phobias of phobic disorders, on the other hand, are characterized primarily by simple avoidance of the anxiety-provoking object or situation.

Obsessions and depression occur together so commonly that discriminating symptom from illness may be difficult. The difficulty is compounded by the fact that obsessionals may develop a depression as florid as that seen in primary affective disorder. During the depression the obsessional symptoms may remain unchanged, worsen, or disappear. Similarly, after the depression the obsessions may be worse, better, or unchanged (the last being the most common outcome) (14).

Episodes of primary depression, in a substantial number of cases, are accompanied by obsessions (14). These are usually ruminative in nature, characterized by guilt and self-deprecation, and mild by comparison with the depressed mood. In more atypical depressions, however, obsessions may dominate the picture. One-third of primary depressives have obsessional personality traits premorbidly and during remissions (33). Obsessional traits apparently precede primary depression as often as they precede obsessional disorder (33).

Obsessional disorder in some instances may be as cyclical as primary depression, with alternating remissions and exacerbations having no apparent relation to life events (21, 34, 41). The history may suggest affective disorder, the symptoms obsessional disorder. Some clinicians treat patients with cyclical obsessions as if they had primary depressions, that is, with antidepressant drugs or electrotherapy. Except for the antidepressant chlorimipramine, discussed later, controlled studies of the results are not available.

In distinguishing between obsessional disorder and primary depression, the following points may be helpful: Compared with primary depressives who have obsessional symptoms, obsessionals who develop a depression do so at an earlier age, have more depressive episodes in their history, exhibit obsessional symptoms episodically during each depression, have a lower rate of attempted suicide, and lack a history of mania (14).

Family history is of diagnostic value when it includes clear-cut depressions, mania, suicide, or alcoholism, which are found more regularly in the families of primary depressives than in the families of patients with any other illness (57).

Mistaking obsessional disorder for schizophrenia is not unusual, especially in the early stages of the illness. Error may arise from difficulty in distinguishing obsessions from delusions and from equating bizarreness or disablement with schizophrenia. Early onset and insidious development are common in both illnesses.

Schizophrenia is characterized by delusions, hallucinations, and *formal* thought disorder (form referring to the flow and connections of thought). Obsessional disorder is primarily a disorder of thought *content*. The speech of obsessionals is understandable; their ideas are queer. The fact that obsessionals recognize they are queer is one of the chief distinctions between obsessional neurosis and schizophrenia.

According to one study, obsessions occur in schizophrenia in about 3 percent of cases, usually early in the course and almost always in the paranoid type (50). When obsessions and schizophrenia symptoms occur together, schizophrenia is usually the most appropriate diagnosis.

Obsessions have been observed in the following organic conditions: encephalitis lethargia—especially during oculogyric crises; early stages of arteriosclerotic dementia; post-traumatic and post-encephalitic states; hearing loss with tinnitus; hypothyroidism (16).

Each of these conditions may be accompanied by the experience that one's mind is working independently, that it is not an integrated part of oneself (33), and by repetitive behavior resembling compulsions. However, the "forced thinking" and "organic orderliness" (15) occasionally observed in brain-damaged patients are said to be less well organized than those that occur in obsessional disorder, and they also lack the sense of internal compulsion that distinguishes the latter condition.

The hallmarks of organic brain disease—confusion, disorientation, memory loss—are not seen in obsessional disorder. Their presence makes the diagnosis comparatively simple. Past history; neurological examination; blood, urine, and spinal fluid studies; plus special imaging techniques help establish the specific cause.

Clinical Management

The data on obsessional disorder justify a certain measure of optimism about its natural course. Spontaneous improvement often occurs, and the patient can be informed of this. She can be reassured that her impulses to commit injury or socially embarrassing acts almost certainly will not be carried out and that she will not—as she often fears—lose her mind. If she needs to be hospitalized, she can be assured that the hospitalization is unlikely to be a long one.

Obsessionals are rarely helped by psychotherapy alone. "Insight" therapy may even be contraindicated. "A searching, interpretive, in-depth approach," wrote one therapist, "in many instances facilitates an introspective obsessive stance" (47). Pessimism about psychotherapy for obsessional disorder is reflected by the scarcity of publications on the subject. Of several hundred articles listed in *Index Medicus* that pertain to therapies for obsessional disorder, only 16 percent are studies of psychotherapy. By far the largest number are studies of behavioral treatments.

Behavior therapy derives from learning theory. According to learning theory, obsessional thoughts are conditioned responses to anxiety-provoking stimuli. Compulsions are established when the individual discovers that the compulsive act reduces the anxiety attendant on obsessional thought. The reduction in anxiety reinforces the compulsive act.

The techniques of behavior therapy are manifold, but their application to obsessional disorder usually involves a single principle: The patient must be exposed to the fear-inducing stimulus. After repeated exposure, the fear disappears because, at bottom, it is unfounded. As fear subsides, so do the obsessions and compulsions. The various techniques for achieving this goal bear such names as desensitization, flooding, implosion, paradoxical intention, operant shaping, and cognitive rehearsal (7, 8, 12).

More than a dozen studies indicate that behavior therapy produces relief from compulsive rituals in many patients; the relief is maintained for at least two to three years following treatment

(36, 38). There is less evidence that obsessions are relieved by behavior therapy. The type of behavior therapy producing the best
results is called "in vivo exposure." "In vivo" means the patients
repeatedly confront the actual source of the anxiety that evokes
the ritual. For example, compulsive handwashers might be asked
to touch the sole of a dirty shoe and otherwise force themselves
to handle dirty objects. The compulsive tidier and checker of
windows might be asked to untidy his or her possessions and deliberately refrain from checking rituals. In many patients, the
anxiety-evoking stimulus eventually will be tolerated without evok-
ing rituals.

Exposure therapy takes up to thirty sessions in the therapist's
office plus self-exposure homework, sometimes with relatives co-
operating as exposure cotherapists. Some patients are too apprehensive to submit to such an ordeal. Others are too depressed. But
as Marks points out in an excellent review (38), the evidence
supporting the efficacy of exposure therapy for compulsive rituals
is impressive.

Many drugs also have been used in the treatment of obsessional
disorder. They include phenothiazines, monoamine oxidase inhibitors (MAOI), lysergic acid diethylamide (LSD), L-tryptophan,
and tricyclic antidepressants (47). With the exception of chlorimipramine, a tricyclic antidepressant, the only evidence for efficacy of these agents has come from single case reports or open
trials. Seven uncontrolled studies (11, 37, 41, 47, 56, 58, 59) and
six double-blind, placebo-controlled studies (2, 4, 23, 37, 38, 53)
of chlorimipramine have indicated autiobsessional effects from
this drug. There has been only one negative report (47).

At this writing, chlorimipramine has not been released for use
in the United States, but it is clearly the drug that shows the
most promise as a specific antiobsessional agent. There has been
much speculation about its mode of action. Chlorimipramine in-
hibits serotonin reuptake and thus potentiates serotonergic activity,
the latter believed to have a mood-altering effect. Tryptophan, the
amino acid precursor of serotonin, is reported to relieve obsessional
symptoms (3, 47), supporting the idea that serotonin in some way

is related to obsessional thinking. Further evidence that inhibition of serotonin reuptake contributes to the efficacy of chlorimipramine comes from an unpublished study in Canada in which patients were treated with fluoxetine, another serotonin reuptake inhibitor (29). A significant improvement was seen. Still another serotonergic agent, zimelidine, was also reported to improve obsessional compulsive disorders, but it was removed from the market because of toxicity (29).

Monoamine oxidase inhibitors (MAOIs) *are* available in the United States, and single case reports with phenelzine sulfate (25) and tranylcypromine sulfate (27) describe marked improvement in obsessional patients. Clorgyline, in a double-blind comparison with chlorimipramine, proved ineffective (23). In other anxiety disorders such as phobic states, MAOIs have been found to be effective (55), so these drugs still show promise.

Benzodiazepines relieve anxiety accompanying obsessions, and antidepressants relieve depression seen with the disorder. Except for chlorimipramine, however, neither group seems to reduce the obsessional thinking.

In summary, the literature indicates that in vivo exposure is the most effective treatment for compulsive rituals and that chlorimipramine is a promising drug for obsessional ideation. Behavior therapy has one strong advantage over drugs: The improvement persists after the treatment ends. Studies agree that chlorimipramine is effective only while the drug is being taken.

A word about electroshock therapy (EST) and lobotomy. The former is viewed by much of the general public and some physicians as a more drastic treatment than, say, antidepressant drugs, but this is arguable. Lobotomy has fallen into disrepute, more justifiably. In any case, there has been only one study of EST for obsessive compulsive disorder.

The study suggests that EST is less effective for depression associated with obsessional disorder than for primary depression (46). In this series, half the obsessionals receiving electroconvulsive therapy (ECT) improved and half did not.

Four follow-up studies (17, 30, 41, 54) indicate that lobotomy produces symptomatic improvement in obsessional disorder superior to that which occurs spontaneously. The more "typical" the obsessional illness, one investigator found, the more likely this was true (30). For a small minority of patients, therefore, lobotomy may warrant consideration. Before undertaking such irreversible treatment, evaluation by two or more psychiatrists is highly desirable. There is general agreement that, if used at all, lobotomy should be reserved for those severely ill obsessional patients with classical symptoms (especially rituals) who fail to respond to other treatments and are totally disabled by their illness.

References

1. Akhtar, S., Wig, N., Varma, V., Pershad, D., and Verma, S. A phenomenological analysis of symptoms in obsessive-compulsive neuroesis. Brit. J. Psychiat. 127:342–348, 1975.

2. Ananth, J., Pecknold, J. C., Van Den Steen, M., and Engelsmann, F. Double-blind comparative study of clorimipramine and amitriptyline in obsessive neurosis. Prog. Neuropsychopharmacol. 5:257–262, 1981.

3. Ananth, J. Clorimipramine in obsessive-compulsive disorder: a review. Psychosomatics. 24:723–727, 1983.

4. Ananth, J. Clorimipramine: an antiobsessive drug. Can. J. Psychiat. 3: 253–258, 1986.

5. Balslev-Olesen, T., and Geert-Jorgensen, E. The prognosis of obsessive-compulsive neurosis. Acta Psychiat. Scand. 34:232–241, 1959.

6. Brown, F. W. Heredity in the psychoneuroses. Proc. R. Soc. Med. 35: 785–790, 1942.

7. Carney, R. M. Behavior therapy and the anxiety disorders: some conceptual and methodological issues. Psychiatr. Dev. 3:65–81, 1985.

8. Cobb, J. Behaviour therapy in phobic and obsessional disorders. Psychiatr. Dev. 1:351–365, 1983.

9. Coryell, W. Obsessive-compulsive disorder and primary unipolar depression: comparisons of background, family history, course, and mortality. J. Nerv. Ment. Dis. 169:220–224, 1981.

10. Dowson, J. The phenomenology of severe obsessive-compulsive neurosis. Brit. J. Psychiat. 131:75–78, 1977.

11. Fernandez, C. E., and Lopez-Ibor Alino, J. La Monochlorimipramina en enfermos psiquiatricors resistentes a otros tratamientos. Acta Luso Esp. Neurol. Psiquiatr. 26:119–147, 1967.

12. Foa, E. B., Steketee, G. S., Ozarow, B. J. Behavior therapy with obsessive-compulsives: from theory to treatment. In *Obsessive-Compulsive*

Disorders: Psychological and Pharmacological Treatments, Mavissakalian, M. (ed.), 49–129. New York: Plenum, 1985.
13. Freud, S.   Notes upon a case of obsessional neurosis. In Standard Edition of the Complete Psychological Works of Sigmund Freud, 10:153. London: Hogarth Press, 1955.
14. Gittleson, N. L.   The phenomenology of obsessions in depressive psychosis. Brit. J. Psychiat. 112:261–264, 1966.
15. Goldstein, K.   After Effects of Brain Injuries in War. Their Evaluation and Treatment. New York: Grune & Stratton, 1942.
16. Goodwin, D. W., Guze, S. B., and Robins, E.   Follow-up studies in obsessional neurosis. Arch. Gen. Psychiat. 20:182–187, 1969.
17. Grimshaw, L.   The outcome of obsessional disorder: a follow-up study of 100 cases. Brit. J. Psychiat. 111:1051–1056, 1965.
18. Hastings, D. W.   Follow-up results in psychiatric illness. Am. J. Psychiat. 114:1057–1065, 1958.
19. Hoover, C. F., Insel, T. R.   Families of origin in obsessive-compulsive disorder. J. Nerv. Ment. Dis. 172:207–215, 1984.
20. Hunter, R., and Macalpine, I.   Three Hundred Years of Psychiatry 1535–1860. London: Oxford Univ. Press, 1963.
21. Ingram, I. M.   Obsessional illness in mental hospital patients. J. Ment. Sci. 107:382–402, 1961.
22. Inouye, E.   Similar and dissimilar manifestations of obsessive-compulsive neurosis in monozygotic twins. Am. J. Psychiat. 121:1171–1175, 1965.
23. Insel, T. R., Murphy, D. L., Cohen, R. M., et al.   Obsessive-compulsive disorder: a double-blind trial of chlorimipramine and clorgyline. Arch. Gen. Psychiat. 40:605–612, 1983.
24. Insel, T. R.   Obsessive-compulsive disorder. Psychiat. Clin. North Am. 8:105–117, 1985.
25. Jain, V. K., Swinson, R. P., Thomas, J. E.   Phenelzine in obsession neurosis Brit. J. Psychiat. 117:237–238, 1970.
26. Jaspers, K.   General Psychopathology. Chicago: Univ. of Chicago Press, 1963.
27. Jenike, M. A.   Rapid response of severe obsessive-compulsive disorder to tranylcypromine. Am. J. Psychiat. 138:1249–1250, 1981.
28. Jenike, M.   Obsessive compulsive disorder: a question of a neurologic lesion. Compr. Psychiat. 25:298–304, 1984.
29. Kahn, R. S., Westenberg, H. G. M., Jolles, J.   Zimelidine treatment of obsessive-compulsive disorder. Acta Psychiat. Scand. 69:259–261, 1984.
30. Kringlen, E.   Obsessional neurotics: a long-term follow-up. Brit. J. Psychiat. 111:709–722, 1965.
31. Langfeldt, G.   Studier av Tvangsfernomenenes forelomist, genese, klinik og prognose. Norsk Laegeforen 13:822–850, 1938.
32. Leckman, J. F., Cohen, D. J.   Recent advances in Gilles de la Tourette syndrome: implications for clinical practice and future research. Psychiat. Dev. 1:301–316, 1983.
33. Lewis, A. J.   Problems of obsessional illness. Proc. R. Soc. Med. 29:325–336, 1936.

34. Lo, W. H. A follow-up study of obsessional neurotics in Hong Kong Chinese. Brit. J. Psychiat. 113:823–832, 1967.
35. Luff, M. C., and Garrod, M. The after results of psychotherapy in 500 adult cases. Brit. Med. J. 11:54–59, 1935.
36. Marks, I., Hodgson, R., and Rachman, S. Treatment of chronic obsessive-compulsive neurosis by in-vivo exposure. Brit. J. Psychiat. 127:349–364, 1975.
37. Marks, I. M., Stern, R. S., Mawson, D., et al. Chlorimipramine and exposure for obsessive-compulsive rituals: I and II. Brit. J. Psychiat. 136: 1–25, 161–166, 1980.
38. Marks, I. M. Review of behavioral psychotherapy, I: obsessive-compulsive disorders. Am. J. Psychiat. 138:584–592, 1981.
39. McGuffin, P., and Mawson, D. Obsessive-compulsive neurosis: two identical twin pairs. Brit. J. Psychiat. 137:285–287, 1980.
40. Moore, J. Of Religious Melancholy, published by Her Majesty's special command. London, 1692.
41. Pollitt, J. Natural history of obsessional states: a study of 150 cases. Brit. Med. J. 1:194–198, 1957.
42. Rasmussen, S. A., Tsuang, M. T. The epidemiology of obsessive compulsive disorder. J. Clin. Psychiat. 45:450–457, 1984.
43. Rennie, T. A. C. Prognosis in the Psychoneurosis: Benign and Malignant Developments, Current Problems in Psychiatric Diagnosis, 66–79. New York: Grune & Stratton, 1953.
44. Robins, L. N., Helzer, J. E., Weissman, M. M., Orvaschel, H., Gruenberg, E., Burke, J. D., and Regier, D. A. Lifetime prevalence of specific psychiatric disorders in three sites. Arch. Gen. Psychiat. 41:949–967, 1984.
45. Rosenberg, C. M. Familial aspects of obsessional neurosis. Brit. J. Psychiat. 113:405–413, 1967.
46. Rüdin, G. Ein Beitrag zur Frage der Zwangskrankheit, insbesondere ihrer heriditaren Beziehungen. Arch. Psychiat. Nervenkr. 191:14–54, 1953.
47. Salzman, L., and Thaler, F. H. Obsessive-compulsive disorders: a review of the literature. Am. J. Psychiat. 138:286–296, 1981.
48. Schilder, P. "Depersonalization." In Introduction to Psychoanalytic Psychiatry. Nerv. Ment. Dis. Monogr., Series 50, 1928.
49. Skoog, G. Onset of anancastic conditions. Acta Psychiat. Scand. 41, Suppl. 184:131, 1965.
50. Stengel, E. A study of some clinical aspects of the relationship between obsessional neurosis and psychotic reaction types. J. Ment. Sci. 91:129, 1945.
51. Stern, R., and Cobb, J. Phenomenology of obsessive-compulsive neurosis. Brit. J. Psychiat. 132:233–239, 1978.
52. Taylor, J. Ductor Dubitantium, or the Rule of Conscience. London: Royston, 1660.
53. Thoren, P., Asberg, M., Cronholm, B., et al. Clorimipramine treatment of obsessive compulsive disorder: a controlled clinical trial. Arch. Gen. Psychiat. 37:1281–1289, 1980.

54. Tippin, J., and Henn, F. Modified leukotomy in the treatment of intractable obsessional neurosis. Am. J. Psychiat. 139:1601–1603, 1982.
55. Tyrer, P., Candy, J., Kelly, D. Phenelzine in phobic anxiety: a controlled trial. Psychol. Med. 3:120–124, 1973.
56. Waxman, D. A general practitioner's investigation on the use of chlorimipramine in the obsessional and phobic disorders. An interim report. J. Int. Med. Res. 1:417–420, 1973.
57. Winokur, G., Clayton, P. J., and Reich, T. *Manic Depressive Illness*. St. Louis: C. V. Mosby, 1969.
58. Yaryura-Tobias, J., and Neziroglu, F. The action of chlorimipramine in obsessive-compulsive neurosis: a pilot study. Curr. Thera. Res. 17:111–116, 1975.
59. Yaryura-Tobias, J., Neziroglu, F., and Bergman, L. Chlorimipramine, for obsessive-compulsive neurosis: an organic approach. Curr. Thera. Res. 20: 541–548, 1976.

6. Phobic Disorders

Definition

A phobia is an intense, recurrent, unreasonable fear of a specific object, activity, or situation that results in a compelling desire to avoid the dreaded object, activity, or situation. A phobic disorder is a chronic condition dominated by one or more phobias. DSM-III-R subdivides phobias into simple phobias (e.g., fear of animals), social phobias (e.g., fear of public speaking), and agoraphobia. The latter is characterized by multiple phobias involving a fear of being in places where help might not be available in the event of a panic attack. DSM-III-R classifies agoraphobia as a subtype of panic disorder, but the condition is still widely regarded as a phobic disorder in which phobias dominate the clinical picture.

Historical Background

Phobos was a Greek god who frightened one's enemies. His likeness was painted on masks and shields for this purpose. "Phobos," or "phobia," came to mean fear or panic (31).

"Phobia" first appeared in medical terminology in Rome two thousand years ago, when hydrophobia was used to describe a symptom of rabies. Though the term was not used in a psychiatric sense until the nineteenth century, phobic fears and behavior were described in medical literature long before that. Hippocrates described at least two phobic persons. One was "beset by terror" whenever he heard a flute, whereas the other could not go beside "even the shallowest ditch," yet could walk in the ditch itself (23).

Robert Burton in *The Anatomy of Melancholy* distinguishes

"morbid fears" from "normal" fears (9). Demosthenes' stage fright was normal, whereas Caesar's fear of sitting in the dark was morbid. Burton believed that morbid fears had little connection with willpower.

The term "phobia" appeared increasingly in descriptions of morbid fears during the nineteenth century, beginning with syphiliphobia, defined in a medical dictionary published in 1848 as "a morbid dread of syphilis giving rise to fancied symptoms of the disease." Numerous theories were advanced to explain phobias, including poor upbringing (31).

In 1871 Westphal described three men who feared public places and labeled the condition agoraphobia, "agora" coming from the Greek word for place of assembly or marketplace (74). Excellent review articles on the historical development of the concept of phobia have been written by Errera (14) and Marks (31). They give Westphal credit for describing phobia in terms of a syndrome rather than an isolated symptom. Westphal even prescribed a treatment for the condition, suggesting that companionship, alcohol, or the use of a cane would be helpful.

Later authorities compiled long lists of phobias, naming each in Greek or Latin terms after the object or situation feared. Thus, as Nemiah points out, "the patient who was spared the pangs of taphaphobia (fear of being buried alive) or ailurophobia (fear of cats) might yet fall prey to belonophobia (fear of needles), siderodromophobia (fear of railways), or triskaidekaphobia (fear of thirteen at table), and pantaphobia was the diagnostic fate of that unfortunate soul who feared them all" (49).

Since the late nineteenth century there has been a continuing controversy over the relationship of phobias to other psychiatric disorders. Janet and Kraepelin, for example, sometimes spoke of phobias and obsessions as though they were synonymous. Others, such as Henry Maudsley and Melanie Klein, considered phobias to be a manifestation of affective illness (33, 65).

As noted, DSM-III-R classifies agoraphobia as a subtype of panic disorder. Regardless of the controversy, clinicians agree that phobias may occur in a variety of phychiatric conditions, but particular types of phobias are the primary manifestation of specific phobic disorders.

Epidemiology

According to a recent survey sponsored by the National Institute of Mental Health (NIMH) (72), roughly one adult in twenty suffers from the most serious variety of phobia, agoraphobia, and one in nine adults harbors some kind of phobia.

The prevalence of agoraphobia and social phobia in two large series of psychiatric patients with mood disorders has been reported to range from 50 to 65 percent (57, 48), and psychiatrists see large numbers of patients with mood disorders. One of the studies (57) had a control group consisting of patients from a fracture clinic in which the prevalence of agoraphobia and social phobia was found to be 30 and 16 percent respectively. However, these figures may have been spuriously elevated owing to the frequency of alcohol abuse in a fracture clinic population. In one study, alcoholics had a high rate of phobia, one-third suffering from disabling phobias and another third from milder phobias (36). Other studies (52, 64, 67) have confirmed an association of alcoholism with social phobias and agoraphobia, but some investigators have failed to find this association (45).

There are two reasons for the discrepancy in prevalence data about phobias: (a) some physicians may include what others consider normal fears in their statistics and (b) people tend to be secretive about phobias. Many people are embarrassed about their phobias and keep them secret. It is not unusual in psychiatric practice to see patients for long periods and then have them describe, almost in passing, a phobia that has plagued them for years.

Clinical Picture

Phobias can be distinguished from "normal" fears by their intensity, duration, irrationality, and the disablement resulting from avoidance of the feared situation.

Table 6.1 Diagnostic criteria for Simple Phobia (DSM-III-R)

A. A persistent fear of a circumscribed stimulus (object or situation) other than fear of having a panic attack (as in Panic Disorder) or of humiliation or embarrassment in certain social situations (as in Social Phobia).

   Note: Do not include fears that are part of Panic Disorder or Agoraphobia.
B. During some phase of the disturbance, exposure to the specific phobic stimulus (or stimuli) almost invariably provokes an immediate anxiety response.
C. The object or situation is avoided, or endured with intense anxiety.
D. The fear or the avoidant behavior significantly interferes with the person's normal routine or with usual social activities or relationships with others, or there is marked distress about having the fear.
E. The person recognizes that his or her fear is excessive or unreasonable.
F. The phobic stimulus is unrelated to the content of the obsessions of Obsessive Compulsive Disorder or the trauma of Post-traumatic Stress Disorder.

DSM-III-R divides phobic disorders into three types: simple, social, and agoraphobia (Tables 6.1, 6.2, 6.3).

A *simple phobia* is an isolated fear of a single object or situation, leading to avoidance of the object or situation. The fear is irrational and excessive but not always disabling because the object or situation can sometimes be easily avoided (e.g., snakes, if you are a city dweller). Impairment may be considerable if the phobic object is common and cannot be avoided, such as a fear of elevators in someone who must use elevators at work.

A person with a simple phobia typically is no more (or less) anxious than anyone else until exposed to the phobic object or situation. Then he or she becomes overwhelmingly uncomfortable and fearful, sometimes having symptoms associated with a panic attack (palpitations, sweating, dizziness, difficulty breathing). The phobic person also fears the *possibility* of confronting the phobic

Table 6.2 Diagnostic criteria for Social Phobia (DSM-III-R)

A. A persistent fear of one or more situations (the social phobic sit-
    uations) in which the person is exposed to possible scrutiny by
    others and fears that he or she may do something or act in a way
    that will be humiliating or embarrassing. Examples include: be-
    ing unable to continue talking while speaking in public, choking
    on food when eating in front of others, being unable to urinate
    in a public lavatory, hand-trembling when writing in the presence
    of others, and saying foolish things or not being able to answer
    questions in social situations.
B. If another disorder is present, the fear in A is unrelated to it, e.g.,
    the fear is not of having a panic attack (Panic Disorder), stutter-
    ing, trembling (Parkinson's disease), or exhibiting abnormal eat-
    ing behavior (Anorexia Nervosa or Bulimia Nervosa).
C. During some phase of the disturbance, exposure to the specific
    phobic stimulus (or stimuli) almost invariably provokes an im-
    mediate anxiety response.
D. The phobic situation(s) is avoided, or is endured with intense
    anxiety.
E. The avoidant behavior interferes with occupational functioning
    or with usual social activities or relationships with others, or there
    is marked distress about having the fear.
F. The person recognizes that his or her fear is excessive or unrea-
    sonable.

---

stimulus. Called anticipatory anxiety, this leads to avoidance of
situations in which the stimulus might be present.

Some common simple phobias are fear of animals, heights,
closed spaces, wind, storms, lightning, loud noises, driving a car,
flying in airplanes, riding in subways, hypodermic needles, and
blood. Less common ones include fear of running water, swallowing solid food, and going to the hairdresser. There is even the case
of the tennis player who wore gloves because he was phobic of
fuzz, and tennis balls are fuzzy. Phobias can develop toward almost any object or situation.

Animal phobias are perhaps the most common simple phobias
or at least the most commonly studied. They occur often in childhood and are usually transient. Among adults, women are subject

Table 6.3 Diagnostic criteria for Agoraphobia (DSM-III-R)

A. Fear of being in places or situations from which escape might be difficult (or embarrassing) or in which help might not be available in the event of suddenly developing a symptom(s) that could be incapacitating or extremely embarrassing. Examples include: dizziness or falling, depersonalization or derealization, loss of bladder or bowel control, vomiting, or cardiac distress. As a result of this fear, the person either restricts travel or needs a companion when away from home, or else endures agoraphobic situations despite intense anxiety. Common agoraphobic situations include being outside the home alone, being in a crowd or standing in a line, being on a bridge, and traveling in a bus, train, or car.

B. Has never met the criteria for Panic Disorder.

Note: If the patient meets criteria for both Agoraphobia and Panic Disorder, the diagnosis should be Panic Disorder with Agoraphobia [see chap. 3].

---

to animal phobias more often than men. The phobia usually involves only one kind of animal but may lead to frequent distress with mild social disablement if the animals are domestic such as dogs or cats (20).

Fear of heights is another common phobia. The fear may be totally unrealistic. The person may be close enough to the ground, so that a fall would not cause injury. Some phobics will not walk down a flight of stairs if they see the open stairwell. Some will not look out a window from the second floor or above, particularly if the window goes from floor to ceiling. Others will not cross a bridge on foot, though they may go by car.

A fear of heights is really a fear of falling—and a fear of falling is really a fear of loss of support (31). More exactly, it is fear of loss of *visual* support. Hence, the person looking out a window experiences no fear if the window is waist-high. A car offers the same protection.

Behind the fear of losing visual support there seems to be a fear of being *drawn* over the edge of the height. The high-up window, the car's doors, provide a sense of protection that no amount of reasoning with oneself can give. The fear of losing support need not even involve heights.

While running for a bus, a woman suddenly felt dizzy and had to hold on to a lamppost for support. This recurred and gradually the patient became terrified to walk anywhere without holding on to a wall or furniture to support her—she was "furniture bound" (31).

Social phobias are another form of isolated phobia that generally lead to only mild forms of impairment. These include fear of eating in public, public speaking, urinating in the presence of others, or other behaviors that may be embarrassing but do not generally lead to total avoidance of the situations. Occasionally, severe handicap may result.

The young wife of a business executive refused to entertain associates of her husband with a dinner party or attend dinner parties in restaurants or someone else's home. Her husband made excuses for her but felt his career was being damaged by their lack of social life. At first her explanation for refusing to join in dinner parties was that it was a waste of time or too much trouble. Later she confessed, tearfully and with much embarrassment, that she was afraid of being *unable* to eat if strangers were watching. Questioned by her concerned husband, she said she was afraid that once food was in her mouth she would be unable to swallow it and then find herself in the embarrassing situation of not knowing where to dispose of the food. She was also afraid that she would gag on the food and possibly vomit.

She could eat in front of her husband with no difficulty until she told him about the phobia and then became concerned that *he* was watching her eat and expecting her to gag or vomit. From then on she ate alone in the kitchen. The marriage survived, but barely (20).

Other social phobias include a fear of practicing musical instruments because the neighbors will hear mistakes; a fear of swimming or undressing in front of others because of shame at one's appearance; and a fear of criticism from superiors.

Unlike simple and social phobias, *agoraphobia* has a complex clinical picture involving various combinations of symptoms, including phobias for the following:

(a) *Public transportation*: trains, buses, subways, airplanes. When crowded, these vehicles may be intolerable to agoraphobics. Waiting in line is almost as bad, whether for a bus or a movie.

(b) *Other confining places*: tunnels, bridges, elevators, the hairdresser or barber's chair, and the dentist's chair. These fears belong to the category of claustrophobia, but most people with claustrophobia have only a single phobia and are not agoraphobics. Paradoxically, agoraphobics may also be fearful of open spaces such as empty parking lots.

(c) *Being home alone*. Some agoraphobics require constant companionship, to the despair of friends, neighbors, and family.

(d) *Being away from home* or another "safe" place where help cannot be readily obtained if needed. The agoraphobic is sometimes comforted just knowing there is a police officer or a doctor somewhere nearby.

In recent years, largely because of observations made by Klein and associates (27, 37), clinicians have come to realize that most agoraphobics give a history of panic attacks preceding the onset of phobias. This has led to the conceptualization of agoraphobia as a complication of panic disorder. According to Klein, agoraphobics experience one or more *spontaneous panic attacks* that lead to *anticipatory anxiety* that in turn leads to *phobic avoidance*. The latter explains why the "housebound housewife" becomes housebound: far better to experience a panic attack in the relative safety of one's home than in places or situations from which escape might be difficult (or embarrassing) or in which help might not be available. A bridge phobia may not involve a fear of bridges per se. Instead, it may reflect a fear of experiencing panic on a bridge. Phobias about buses, trains, or airplanes may arise from concern that a panic attack might occur in a vehicle from which escape is impossible or embarrassing.

Klein's concept

panic attack ⟶ anticipatory anxiety ⟶ phobic avoidance

is a concept now shared by many clinicians that has led to a departure from conventional nomenclature in DSM-III-R. Instead of being listed under phobic disorders, as occurred in previous manuals, DSM-III-R adopted Klein's view of the disorder and described agoraphobia as a subtype of panic disorder. Depending on which symptoms predominate—the panic attacks or phobias—agoraphobia may fall in one of three categories: Panic Disorder with Agoraphobia, Panic Disorder without Agoraphobia, and Agoraphobia without History of Panic Disorder. (The term "panic disorder" first received official recognition in DSM-III, replacing the term anxiety neurosis and a host of synonyms for this term [see chap. 3].)

The identification of panic disorder as a condition often associated with agoraphobia surprised clinicians accustomed to viewing panic disorder (anxiety neurosis) as a condition that is more common than agoraphobia and one not requiring the qualifying phrase "without agoraphobia." Nosological controversies, however, are traditional in psychiatry.

Klein concedes that agoraphobics do not always give a history of panic disorder. However, he maintains that even those who do not give this history still experience "limited symptom attacks," which he views as milder forms of panic (27). Architects of DSM-III-R decided to define panic attacks as requiring four or more of thirteen symptoms and "limited symptom attacks" as involving fewer than four (Table 3.2). In an interview (27), Klein gave an example of what he meant by "limited symptom attacks," citing "patients with sudden dizzy feelings who feel as though their heads are floating off their shoulders and that they might fall down." Such patients may not experience a fully developed panic attack, involving cardiorespiratory symptoms. Some symptoms seen in "limited symptom attacks" may lead to confusion of anxiety states with Meniere's disease. From personal experience the essayist, E. B. White described the symptoms humorously (12):

> I will be walking along the street, say, and will take three normal steps in a forward direction: then, as I am about to set my foot down for the fourth step, the pavement moves an inch or two to the right and drops off three-quarters of an inch, and I am not quick enough for it. This results in my jostling somebody on my left, or hitting the corner of the Fred F. French building a glancing blow. It was fun for a

few days, but I have recovered from the first fine ecstasy of dizziness, and am getting bored with it. Once I sidled into a police horse, and he gave me back as good as I gave him.

White saw a doctor who "partly by stealth, partly by cunning," gained entrance into his middle ear, hoping to discover there the secret of the dizziness.

Agoraphobics have other psychiatric symptoms than panic attacks and phobias. They tend to be anxious people. They are subject to depression and often experience an eerie feeling called "depersonalization." The feeling is hard to define and appears to be an unpleasant, scary sense of being unreal, strange, and disembodied, cut off from one's surroundings.

Agoraphobia is the most disabling of the phobic disorders. When severe, it can be as disabling as the most crippling forms of schizophrenia. Although agoraphobics almost never require chronic hospitalization, they are sometimes unable to leave home for months or years at a time. The more extreme cases may confine themselves to a single room or spend most of their time in bed. Even milder cases involve restrictions in social functioning. Victims are unable to visit friends and neighbors or go on family outings. They recruit other people to do their shopping and take their children to school. They postpone seeing the dentist and may cut their own hair. Fortunately, new treatments—behavioral and pharmacological—have provided relief for many (see Clinical Management).

Natural History

In three studies (2, 13, 34) phobic individuals have been followed for long periods to see what happens to them and their phobias. Here are the conclusions:

Simple phobias that begin in childhood and early adolescence tend to improve and eventually disappear. Five years after onset, about half of the young victims are symptom-free and almost all are improved.

Phobias that begin after adolescence continue for longer periods, with about half of patients improved after five years but only about 5 percent symptom-free. In the older group, the phobia gradually becomes more severe in about one-third of patients. About 20 percent are impaired in their work or social functioning because of the phobia.

Gender has little influence on the outcome of simple phobia: Women do as well (or poorly) as men, with age of onset being the most important prognostic indicator.

Simple phobias as a group have a more favorable prognosis than the other phobic disorders. They almost always involve a single isolated phobia, whereas social phobias more often involve two or more phobias and agoraphobia invariably involves multiple phobias.

Most social phobias first occur in adolescence or the early twenties. They rarely begin before puberty or after age thirty. Unlike simple phobias and agoraphobia, which affect women more often than men, social phobias affect both sexes equally (29, 66). They are the commonest type of phobia in men. Most social phobias develop over several months, stabilizing after a period of years with a gradual diminution of severity in middle life. Most begin without any apparent precipitating event.

Agoraphobia usually develops in the middle or late twenties, almost never occurring before eighteen or after thirty-five. Of the three types of phobia, agoraphobia has the latest age of onset; simple phobia occurs most often in childhood and social phobia most often in adolescence (42, 69).

Simple phobias seem more strongly related to the occurrence of a single traumatic event than do social phobias. The onset of agoraphobia is particularly obscure. Agoraphobics rarely associate the onset of their phobias with a single frightening event, but vividly recall their *first* panic attack. Freud reported this first. In agoraphobia, he wrote, "We often find a recollection of a state of panic; and what the patient actually fears is a repetition of such an attack under those special conditions in which he believes he cannot escape it" (18).

Agoraphobics rarely have an explanation for the first anxiety attack. "It came out of the blue," they tend to say. They associate future anxiety attacks with other unpleasant possibilities—a fainting spell, a sudden illness, something embarrassing—but their real reason for staying in the house and avoiding open spaces and public places is to avoid another attack. What makes the prospect of an anxiety attack especially frightening is the unexplained nature of the first attack: If it happened for no obvious reason once, it could happen again for no obvious reason.

The first anxiety attack often occurs against a background of worry and unhappiness: job dissatisfaction, a domestic crisis, a serious medical illness, a death in the family. The relationship of nonspecific stress and the onset of agoraphobia has led some to call agoraphobia the "calamity syndrome." Nevertheless, agoraphobics do not associate the first anxiety attack with a *specific* stress that would justify panic. Isaac Marks quotes a patient as saying (31):

I was standing at the bus stop wondering what to cook for dinner when suddenly I felt panicky and sweaty, my knees felt like jelly, and I had to hold on to the lamppost and I was afraid I would die. I got on the bus and was terribly nervous, but just managed to totter home. Since then I haven't liked to go out on the street and have never been on a bus again.

Marks comments:

In such cases the primary event is anxiety which arose within the patient for no accountable reason. The phobia developed secondarily as the anxiety is attached to the immediate environment prevailing when the internal anxiety happens to surge. First comes the panic without relationship to the surroundings, the surroundings are then attached or "conditioned" to the anxiety and agoraphobia has begun.

One-half to three-quarters of mild or moderately disabled agoraphobics recover or substantially improve five to ten years after the illness begins. The severely disabled housebound patient may remain that way indefinitely. Some cases are short-lived but nobody knows how many. If the symptoms of agoraphobia persist for a year, they are almost certain to last much longer (31).

Complications

Social, occupational, and marital disability are the main complications of phobias (58). In isolated phobias the disability tends to be mild and is limited to responses to the phobic object or situation. Agoraphobics suffer greater impairment, sometimes to the point of being totally housebound.

In several series of patients, frigidity and other sexual maladjustments have been reported in association with agoraphobia. The frigidity often antedates the phobias (55, 71).

Agoraphobics occasionally have an affective syndrome indistinguishable from primary affective disorder except for the prior history of chronic phobias. Suicide does not appear to be a complication of phobic disorder. At least five studies (7, 36, 52, 64, 67) indicate that agoraphobics and people with social phobias are particularly susceptible to alcohol dependence, but two other studies (45, 72) suggest the contrary and the issue is unresolved.

In recent years, much attention has been focused on the relationship between mitral valve prolapse and anxiety disorders. It has been reported that as many as 50 percent of patients with panic disorder have mitral valve prolapse, which is considerably higher than the 4- to 17-percent prevalence reported in general population studies (11, 21, 24, 39). Kantor and associates found that 44 percent of twenty-five agoraphobic women had mitral valve prolapse compared with 17 percent of twenty-three age-matched controls (24).

Family Studies

In an Iowa study (22), 32 percent of first-degree relatives of agoraphobics had an anxiety disorder (the collective term for agoraphobia, panic disorder, generalized anxiety, social phobia, simple phobia, and obsessive compulsive disorder). Of the first-degree relatives, 9 percent had agoraphobia. Of the first-degree relatives of nonanxious controls studied, 15 percent had anxiety disorders and 4 percent had agoraphobia. The study also revealed that 35 percent of the fathers of agoraphobics were alcoholic compared to 10 percent of fathers of controls. Affective disorders were more common in the first-degree relatives of the nonanxious controls (10 percent) than in the families of agoraphobics (7 percent), disputing the view that agoraphobia is a variant of affective disorder. In another study of sixty agoraphobics, 12 percent of the relatives had the illness, a rate significantly higher than in the general population (46).

Simple and social phobia apparently do not run in families (20, 31).

Several studies indicate that phobics come from stable families in which the mothers may be overprotective (55, 71). Of five sets of monozygotic twins studied at the Maudsley Hospital in which one of each twin pair was phobic, there were no cases in which both twins were affected (31).

Differential Diagnosis

Phobias occur in psychiatric conditions other than phobic disorders, notably obsessional disorder and panic disorder. Because all three conditions begin early in life and have a chronic though sometimes fluctuating course, distinguishing them may be difficult. This is particularly true of agoraphobia when multiple phobias occur together with panic attacks, obsessional thinking, and other nonphobic psychiatric symptoms.

Panic disorder patients often have phobias, but the most conspicuous feature of panic disorder is anxiety attacks unrelated to external situations in which the subjective experience of fear is accompanied by cardiorespiratory symptoms. Obsessive compulsive disorder is manifested by a panoply of obsessional symptoms, including checking and counting rituals, repetitive ideas relating to the possibility of harming oneself or others, and other recurrent thoughts.

Like patients with other anxiety disorders, phobics may become depressed and develop the full range of affective symptoms, physiological and subjective. When the phobias preceding the affective episode are isolated and nonincapacitating, it may be legitimate to view the affective disorder as "primary." With regard to agoraphobia, depressions commonly occur during the course of the illness and probably should be considered "secondary" (without giving this nosologic distinction etiological connotations). Some British authors consider agoraphobia an atypical form of affective disorder. Many of the symptoms of primary affective disorder may have phobic qualities such as fearfulness of social situations and disease.

When phobiclike symptoms occur in schizophrenia, they are usually bizarre and unaccompanied by insight.

Clinical Management

Regardless of the theoretical orientation of the therapist, it is generally agreed that at some point the phobic patient must confront the feared situation. Freud observed that "one can hardly ever master a phobia if one waits till the patient lets the analysis influence him to give it up. One succeeds only when one can induce them through the influence of the analysis to go about alone and to struggle with their anxiety while they make the attempt" (18).

Most treatments have been designed to reduce the phobic anxiety to the extent that the patient can tolerate exposure to the phobic situation. These treatments include drugs, psychotherapy, and behavior modification techniques.

A number of double-blind, placebo-controlled studies have demonstrated the efficacy of imipramine hydrochloride (61, 76, 77, 78) and phenelzine sulfate (47, 69) in the treatment of phobic disorders, particularly agoraphobia associated with panic attacks. Uncontrolled clinical observations suggest that other antidepressants may also be effective.

In 1983 two additional large studies of imipramine and phenelzine were published, one by Klein and his colleagues in New York (26, 78) and the other by Marks and his British colleagues (36). They arrived at somewhat different conclusions. The New York group found that imipramine is effective in the treatment of panic attacks in agoraphobic patients; not all agoraphobics, however, have panic attacks. Furthermore, dynamically oriented supportive psychotherapy was reported to be as effective as behavior therapy, as long as both led to confrontation with the phobic situation. In contrast, the British group reported that imipramine was not superior to placebo in preventing panic attacks in the agoraphobic patient and that in vivo exposure (exposure to real-life phobic stimuli) was a specific and potent treatment for agoraphobia. The doses of imipramine and the schedule of administration were different in the two studies, and this may explain the discrepancy.

Imipramine and other tricyclic antidepressants as well as phenelzine and other monoamine oxidase inhibitors (MAOIs) may only be effective in preventing panic attacks when the patient is taking the medication (26, 36); there is some disagreement on this point. Marks et al. (36) found that patients treated with in vivo exposure were much improved six weeks after the drug and behavioral treatment, but there was no benefit for the group treated only with phenelzine after the drug had been discontinued. Klein sometimes finds improvement persisting after the drug is stopped. He also has switched from imipramine to desipramine because the latter has fewer anticholinergic side effects and greater patients compliance (27).

Klein (26) distinguishes between the panic component of agoraphobia and the anticipatory anxiety patients experience between panic attacks. Using higher doses of imipramine than did the British group, Klein reported a high rate of success in preventing panic attacks but no success in relieving anticipatory anxiety. Benzodiazepines have been suggested as an effective treatment for the latter.

Buspirone, the nonbenzodiazepine anxiolytic introduced in late 1986, however, may be the drug of choice for anticipatory anxiety. Unlike the benzodiazepines, which produce mild physical dependence, buspirone (68) apparently produces none and is as effective as the benzodiazepines in relieving anxiety.

In a recent review (29) Liebowitz and his coauthors called attention to several studies in which patients with social phobias have received relief with beta-adrenergic receptor blockers. Antidepressants of the MAOI class have also been recommended for social phobia and a single case report indicates that clonidine may be useful (66). Liebowitz calls social phobia the "neglected anxiety disorder." Treatments for the problem began to emerge in the mid-1980s.

At present, the most popular approach to the treatment of simple and social phobias involves techniques based on learning theory. Of these, "systematic desensitization," introduced by Wolpe, has been most widely used and studied (75). Nearly twenty studies, most of them controlled, indicate that systematic desensitization produces at least temporary relief of isolated phobias but is less effective in agoraphobia (50, 51, 53).

The method consists of training patients to relax and then gradually exposing them to more and more intensive versions of the phobic stimulus. This requires that patients imagine the feared events, starting with events that produce minimal anxiety and then progressing through a "hierarchy" to imagine events of maximal fearfulness. At this point, they are expected to confront the phobic situation in reality and by relaxation be able to tolerate it. The approach is based on the concept of "reciprocal inhibition" that holds that two mutually incompatible emotions such as fear and pleasure cannot be simultaneously experienced. Therefore, relaxation, being "pleasurable," theoretically cancels the sensation of fear if the latter is not overwhelming.

A variation of this approach consists of "flooding" the individual with the phobic stimulus at maximal intensity without progressing through a hierarchy. Flooding is apparently most effective with in vivo exposure. If the person is afraid of heights, the therapist may take him to the roof of a tall building and have him lean over the guardrail. The patient will be terrified. Terror, however, is not an emotion that can be sustained indefinitely and, with continued exposure, he will find his anxiety diminishing.

Real-life flooding is often hard to arrange, so therapists fall back on imaginary flooding. After learning what frightens the patient the most, the therapist asks the patient to visualize his most frightening scene and helps by describing the scene. The therapist watches for signs of anxiety in the patient and repeats the scenes that produce the most anxiety. As with real-life flooding, the therapist keeps it up until the anxiety level falls. A recent comparison of imipramine and flooding in sixty-two agoraphobic patients indicated that imipramine had superior effects (40).

Some therapists embellish flooding by describing exaggerated catastrophic scenes to the patient. The patient who is afraid to drive is asked to imagine that she is driving a school bus that plunges over a cliff with everyone killed. The patient who is frightened of speaking out at a small meeting is told to imagine that she is the keynote speaker at a convention in Madison Square Garden. This is called *implosion therapy*.

What works best? There are studies showing that flooding is better than systematic desensitization; studies showing that systematic desensitization is better than flooding; and studies showing that both have the same effect or no effect (20, 43). What almost all studies show is that real-life exposure is superior to imaginary exposure.

References

1. Agras, S., Sylvester, D., and Oliveau, D. The epidemiology of common fears and phobia. Compr. Psychiat. 10:151–156, 1969.
2. Agras, S., Chapin, H. N., and Oliveau, D. C. The natural history of phobia, course and prognosis. Arch. Gen. Psychiat. 26:315–317, 1972.
3. Ayd, F. J., Jr. Antidepressant, 1959. Psychosomatics 1:31, 1960.
4. Bates, H. D. Predictive power of the reinforcement survey schedule; near null results. Newsl. Res. Psychol. 12:113–116, 1970.
5. Berg, I. School phobia in the children of agoraphobic women. Brit. J. Psychiat. 128:86–89, 1976.
6. Berg, I., Butler, A., and Pritchard, J. Psychiatric illness in the mothers of school phobic adolescents. Brit. J. Psychiat. 125:466–467, 1974.
7. Bowen, R. C., Cipywynk, D., D'Arcy, D., and Keegan, D. Alcoholism, anxiety disorders, and agoraphobia. Alcoholism: Clin. Exp. Res. 8:48–50, 1984.
8. Buglass, D., Clarke, J., Henderson, A. S., Kreitman, N., and Presley, A. S. A study of agoraphobic housewives. Psychol. Med. 7:73–86, 1977.

9. Burton, R. *The Anatomy of Melancholy* (1621), vol. 1, 11th ed. London, 1813.

10. Dally, P. J., and Rohde, P. Comparison of antidepressant drugs in depressive illnesses. Lancet 1:18, 1961.

11. Darsee, J. R., Mikolich, J. R., Nicoloff, N. B., et al. Prevalence of mitral valve prolapse in presumably healthy young men. Circulation 59: 619–622, 1979.

12. Elledge, S. *E. B. White, a Biography.* New York: Oxford Univ. Press, 1984.

13. Emmelkamp, P. M. G., and Kuipers, A. C. M. Agoraphobia: a follow-up study four years after treatment. Brit. J. Psychiat. 134:342–355, 1979.

14. Errera, P. Some historical aspects of the concept of phobia. Psychiat. Q. 36:325–336, 1962.

15. Errera, P., and Coleman, J. V. A long-term follow-up study of neurotic patients in a psychiatric clinic. J. Nerv. Ment. Dis. 136:267, 1963.

16. Frankel, F., and Orne, M. Hypnotizability and phobic behavior. Arch. Gen. Psychiat. 33:1259–1261, 1976.

17. Frazier, S. H., and Carr, A. C. Phobic reactions. In *Comprehensive Textbook of Psychiatry.* Freedman, A. M., and Kaplan, H. E. (eds.). Baltimore: Williams & Wilkins, 1967.

18. Freud, S. *Collected Papers,* vol. 2, New York: Basic Books, 1950.

19. Goldman, D. Clinical experience with newer antidepressant drugs and some related electroencephalographic observations. Ann. N.Y. Acad. Sci. 80:687, 1959.

20. Goodwin, D. W. *Phobia: The Facts.* London: Oxford Univ. Press, 1983.

21. Gorman, J. M., Fyer, A. F., Gliklich, J., et al. Mitral valve prolapse and panic disorder: effect of imipramine. In *Anxiety Revisited,* Klein, D. F. Rabkin, J. G. New York: Raven Press, 1981.

22. Harris, E. L., Noyes, R., Crowe, R. R., and Chaudhry, D. R. A family study of agoraphobia: report of a pilot study. Arch. Gen. Psychiat. 40: 1065–1070, 1983.

23. Hippocrates on Epidemics, V, Section LXXXII.

24. Kantor, J. S., Zitrin, C. M., Zeldis, S. M. Mitral valve prolapse syndrome in agoraphobic patients. Am. J. Psychiat. 127:467–469, 1980.

25. Klein, D. F. The delineation of two drug-responsive anxiety syndromes. Psychopharmacologia 5:397, 1964.

26. Klein, D. F., Zitrin, C. M., Woerner, M. G., and Ross, D. C. Treatment of phobias. II. Behavior therapy and supportive psychotherapy: Are there any specific ingredients? Arch. Gen. Psychiat. 40:139–145, 1983.

27. Klein, D. F. Interview. Currents. 4:5–10, October 1985.

28. Lang, P. J. Fear reduction and fear behaviour. "Problems in treating a construct." In *3d Conference on Research in Psychotherapy,* Chicago, June 1966.

29. Liebowitz, M. R., Gorman, T. M., Fyer, A. T., and Klein, D. F. Social phobia: review of a neglected anxiety disorder. Arch. Gen. Psychiat. 42: 729–736, 1985.

30. Marks, I. M., Birley, J. L. T., and Gelder, M. G. Modified leucotomy in severe agoraphobia: a controlled serial inquiry. Brit. J. Psychiat. 112:757–769, 1966.
31. Marks, I. M. Fears and Phobias. London: Academic Press, 1969.
32. Marks, I. M. Agoraphobic syndrome (phobic anxiety state). Arch. Gen. Psychiat. 23:538–553, 1970.
33. Marks, I. M. The classification of phobic disorders. Brit. J. Psychiat. 116:377–386, 1970.
34. Marks, I. M. Phobic disorders four years after treatment: a prospective follow-up. Brit. J. Psychiat. 118:683–688, 1971.
35. Marks, I. M., Boulougouris, J., and Marset, P. Flooding versus desensitization in the treatment of phobic patients: a crossover study. Brit. J. Psychiat. 119:353–375, 1971.
36. Marks, I. M., Gray, S., Cohen, D., Hill, R., Mawson, D., Ramm, E., and Stern, R. S. Imipramine and brief therapist-aided exposure in agoraphobics having self-exposure homework. Arch. Gen. Psychiat. 40:153–162, 1983.
37. Marks, I. Are there anticompulsive or antiphobic drugs? Review of the evidence. Brit. J. Psychiat. 143:338–347, 1983.
38. Mavissakalian, M., Michelson, L., and Dealy, R. S. Pharmacological treatment of agoraphobia: imipramine versus imipramine with programmed practice. Brit. J. Psychiat. 143:348–355, 1983.
39. Mavissakalian, M., Salerni, R., Thompson, M. E., and Michelson, L. Mitral valve prolapse and agoraphobia. Am. J. Psychiat. 140:1612–1614, 1983.
40. Mavissakalian, M., and Michelson, L. Agoraphobia: relative and combined effectiveness of therapist-assisted in vivo exposure and imipramine. J. Clin. Psychiat. 47:117–122, 1986.
41. Mathews, A., Johnston, D., Lancashire, M., Munby, M., Shaw, P., and Gelder, M. Imaginal flooding and exposure to real phobic situations: treatment outcome with agoraphobic patients. Brit. J. Psychiat. 129:362–371, 1976.
42. Mathews, A. M., Gelder, M. G., and Johnston, D. W. Agoraphobia: Nature and Treatment. New York: Guilford Press, 1981.
43. Mavissakalian, M., and Barlow, D. H. (eds.). Phobia: Psychological and Pharmacological Treatment. New York: Guilford Press, 1981.
44. McConaghy, N. Results of systematic desensitization with phobias re-examined. Brit. J. Psychiat. 117:89–92, 1970.
45. Mendelson, J. H., Miller, K. D., Mello, N. K., Pratt, H., and Schmitz, R. Hospital treatment of alcoholism: a profile of middle income Americans. Alcoholism: Clin. Exp. Res. 6:377–383, 1982.
46. Moran, C., Andrews, G. The familial occurrence of agoraphobia. Brit. J. Psychiat. 146:262–267, 1985.
47. Montjoy, C., Roth, M., Garside, R., and Leitch, I. A clinical trial of phenelzine in anxiety depressive and phobic neuroses. Brit. J. Psychiat. 131:486–492, 1977.

48. Mullaney, J. A., and Trippett, C. J. Alcohol dependence and phobias: clinical description and relevance. Brit. J. Psychiat. 135:565–573, 1979.
49. Nemiah, J. D. Phobic disorder. In Comprehensive Textbook of Psychiatry, 3rd ed. Freedman, A. M., and Kaplan, H. E. (eds.). Baltimore: Williams & Wilkins, 1980.
50. Paul, G. L. Outcome of systematic desensitization—1. background and procedures, and uncontrolled reports of individual treatments. In Behavior Therapy: Appraisal and Status, Franks, C. M. (ed.). New York: McGraw-Hill, 1969.
51. Paul, G. L. Outcome of systematic desensitization—2. controlled investigations of individual treatment, technique variations, and current status. In Behavior Therapy: Appraisal and Status, Franks, C. M. (ed.). New York: McGraw-Hill, 1969.
52. Quitkin, F. M., and Rabkin, J. G. Hidden psychiatric diagnosis in the alcoholic. In Solomon, J. (ed.), 130–131. Alcoholism and Clinical Psychiatry. New York: Plenum, 1982.
53. Rimm, D. C., and Masters, J. C. Behavior Therapy: Techniques and Empirical Findings, 2d ed. New York: Academic Press, 1979.
54. Roberts, A. H. Housebound housewives: a follow-up study of a phobic anxiety state. Brit. J. Psychiat. 110:191–197, 1964.
55. Roth, M. The phobic-anxiety-depersonalisation syndrome. Proc. R. Soc. Med. 52:587, 1959.
56. Sbarbaro, J. A., Mehlman, D. J., Wu, L., et al. A prospective study of mitral valvular prolapse in young men. Chest 75:555–559, 1979.
57. Schapira, K., Kerr, T. A., and Roth, M. Phobias and affective illness. Brit. J. Psychiat. 117:25–32, 1970.
58. Shafar, S. Aspects of phobic illness—a study of 90 personal cases. Brit. J. Med. Psychol. 49:221–236, 1976.
59. Shapiro, M. B., Marks, I. M., and Fox, B. A therapeutic experiment on phobic and affective symptoms in an individual psychiatric patient. Brit. J. Soc. Clin. Psychol. 2:81–93, 1963.
60. Shaw, H. A simple and effective treatment for flight phobia. Brit. J. Psychiat. 130:229–232, 1977.
61. Sheehan, D. V., Ballenger, J., and Jacobsen, G. Treatment of endogenous anxiety with phobic, hysterical, and hypochondriacal symptoms. Arch. Gen. Psychiat. 37:51–59, 1980.
62. Sheehan, D. V., Sheehan, K. E., and Minichiello, W. E. Age of onset of phobic disorders: a reevaluation. Compr. Psychiat. 22:544–553, 1981.
63. Sim, M., and Houghton, H. Phobic anxiety and its treatment. J. Nerv. Ment. Dis. 143:484–491, 1966.
64. Smail, P., Stockwell, T., Canter, S., and Hodgson, R. Alcohol dependence and phobic anxiety states—I. A prevalence study. Brit. J. Psychiat. 144:53–57, 1984.
65. Snaith, R. P. A clinical investigation of phobias. Brit. J. Psychiat. 114:673–698, 1968.
66. Solyom, L., Ledwidge, B., and Solyom, C. Delineating social phobia. Brit. J. Psychiat. 149:464–470, 1986.

67. Stockwell, T., Smail, P., Hodgson, R., and Canter, S. Alcohol dependence and phobic anxiety states—II. A retrospective study. Brit. J. Psychiat. 144:58–63, 1984.
68. Sussman, N. Treatment of anxiety with buspirone. Psychiatr. Ann. 17: 114–120, 1987.
69. Tyrer, P., Candy, J., and Kelly, D. Phenelzine in phobic anxiety: a controlled trial. Psychol. Med. 3:120–124, 1973.
70. Tyrer, P., and Steinberg, D. Symptomatic treatment of agoraphobia and social phobias: a follow-up study. Brit. J. Psychiat. 127:163–168, 1975.
71. Webster, A. S. The development of phobias in married women. Psychol. Monogr. 67:367, 1953.
72. Weissman, M. M., Myers, J. K., Harding, P. S. Prevalence and psychiatric heterogeneity of alcoholism in a United States urban community. J. Stud. Alcohol 41:672–681, 1980.
73. West, E. D., and Dally, P. J. Effects of iproniazid in depressive syndromes. Brit. Med. J. 1:1491, 1959.
74. Westphal, C. Die Agoraphobie: eine neuropathische Erscheinuung. Arch. Psychiat. Nervenk. 3:138–171, 219–221, 1871–72.
75. Wolpe, J. Psychotherapy by Reciprocal Inhibition. Stanford: Stanford Univ. Press, 1958.
76. Zitrin, C. M., Klein, D., and Woerner, M. G. Behavior therapy, supportive psychotherapy, imipramine and phobias. Arch. Gen. Psychiat. 35: 307–316, 1978.
77. Zitrin, C. M., Klein, D. F., and Woerner, M. G. Treatment of agoraphobia with group exposure in vivo and imipramine. Arch. Gen. Psychiat. 37:63–72, 1980.
78. Zitrin, C. M., Klein, D. F., Woerner, M. G., and Ross, D. C. Treatment of phobias. I. Comparison of imipramine hydrochloride and placebo. Arch. Gen. Psychiat. 40:125–138, 1983.

7. Alcoholism

Definition

Alcoholism has been defined by Keller and his associates (68) as the "repetitive intake of alcoholic beverages to a degree that harms the drinker in health or socially or economically, with indication of inability consistently to control the occasion or amount of drinking." A synonym for alcoholism recommended by the World Health Organization (25, 51) and DSM-III-R is *alcohol dependence*.

Historical Background

The use of alcohol by man dates back at least to Paleolithic times. The evidence for this statement derives from etymology as well as studies of Stone Age cultures that survived into the twentieth century.

Available to Paleolithic man, presumably, were fermented fruit juice (wine), fermented grain (beer), and fermented honey (mead). Etymological evidence suggests that mead may have been the earliest beverage of choice. The word "mead" derives— by way of *mede* (Middle English) and *meodu* (Anglo-Saxon)— from ancient words of Indo-European stock, such as *methy* (Greek) and *madhu* (Sanskrit). In Sanskrit and Greek, the term means both "honey" and "intoxicating drink." The association of honey rather than grain or fruit with intoxication may indicate its greater antiquity as a source of alcohol (103).

All but three of the numerous Stone Age cultures that survived into modern times have been familiar with alcohol. "The three exceptions," Berton Roueché writes, "are the environmentally underprivileged polar peoples, the intellectually stunted Australian aborigines, and the comparably lackluster primitives of Tierra del

Fuego" (103). Early European explorers of Africa and the New World invariably discovered that alcohol was important in the local cultures. The Indians of eastern North America, for instance, were using alcohol in the form of fermented birch and sugar maple sap (103).

Alcohol was used medicinally and in religious ceremonies for thousands of years, but it also has a long history of recreational use. Noah, according to the Old Testament, "drank of the wine and was drunken." Mesopotamian civilization provided one of the earliest clinical descriptions of intoxication and one of the first hangover cures. In *The History of Medicine* (109), Sigerist translates a Mesopotamian physician's advice: "If a man has taken strong wine, his head is affected and he forgets his words and his speech becomes confused, his mind wanders and his eyes have a set expression; *to cure him*, take licorice, beans, oleander . . . to be compounded with oil and wine before the approach of the goddess Gula [or sunset], and in the morning before sunrise and before anyone has kissed him, let him take it, and he will recover."

Roueché notes that "one of the few surviving relics of the Seventeenth Egyptian Dynasty, which roughly coincided with the reign of Hammurabi, is a hieroglyphic outburst of a female courtier. 'Give me eighteen bowls of wine!' she exclaims for posterity. 'Behold, I love drunkenness!' " (103). So did other Egyptians of that era. "Drunkenness was apparently not rare," Sigerist remarks, "and seems to have occurred in all layers of society from the farmers to the gods (or ruling class). Banquets frequently ended with the guests, men and women, being sick, and this did not in any way seem shocking" (109).

Not only do descriptions of drunkenness fill the writings of antiquity but also pleas for moderation. Dynastic Egypt apparently invented the first temperance tract (103). Moderation was recommended by no less an authority than Genghis Khan: "A soldier must not get drunk oftener than once a week. It would, of course, be better if he did not get drunk at all, but one should not expect the impossible." The Old Testament condemns drunkenness, but not alcohol. "Give strong drink unto him that is ready to perish,"

the Book of Proverbs proclaims, "and wine unto those that be of
heavy hearts. Let him drink, and forget his poverty, and remember
his misery no more."

The process of distillation was discovered about A.D. 800 in Ara-
bia. ("Alcohol" comes from the arabic *alkuhl,* meaning essence.)
For centuries distilled alcohol was used in medicine, but by the
seventeenth century it had also become a drug of abuse on a large
scale. By the late seventeenth century the annual worldwide production of distilled liquors, chiefly gin, was enormous.

Ancient or classical writers used words that are universally trans-
lated as drunkenness. The people who had it were drunkards. In
the fourteenth century Chaucer used *dronkelewe* to mean addiction to alcohol as a mental illness (14). By the nineteenth century "inebriety" was the favored descriptor; those who manifested
it were not merely inebriated—they were inebriates. A Swedish
public health authority, Magnus Huss, in 1849 coined the term
"alcoholism." The word caught on: Danish has *alkoholisme;* Dutch,
alcoholisme; English, *alcoholism;* Finnish, *alkoholismi;* German,
alkoholismus; Italian, *alcolismo;* Norwegian, *alkoholisme;* Polish,
alkoholizm; Portuguese, *alcoolismo;* Russian, *alkogolism;* Serbo-
Croatian, *alkoholizam;* Slovene, *alkoholizem;* Spanish, *alcohol-
ismo;* Swedish, *alkoholism* (67).

The "disease concept" of alcoholism originated in the writings
of Benjamin Rush and the British physician Thomas Trotter (104),
and during the last half of the nineteenth century the notion that
alcoholism was a disease became popular with physicians. In the
1830s, Dr. Samuel Woodward, the first superintendent of Worces-
ter State Hospital, Massachusetts, and Dr. Eli Todd of Hartford,
Connecticut, suggested establishing special institutions for inebriates. The first was opened in Boston in 1841. In 1904 the Medical
Temperance Society changed its name to the American Medical
Association for the Study of Inebriety and Narcotics. The *Journal
of Inebriety,* established in 1876, was founded on the "fact that
inebriety is a neurosis and psychosis." During Prohibition, however, the concept of alcoholism as a disease lost its vogue (62).

With repeal of the Eighteenth Amendment, the disease concept was revived. Pioneering studies performed at the Yale School of Alcohol Studies and the writings of E. M. Jellinek were largely responsible for the popularization of this concept in the twentieth century. In the mid-1960s the U.S. Government began supporting alcoholism research on a rather large scale; by the 1980s the federal government and most state governments were sponsoring treatment programs for alcoholics, and a large number of proprietary hospitals for alcoholism spanned the continent.

Epidemiology

More is known about patterns of "normal" drinking than about the prevalence of alcoholism. In 1985 the federal government studied the drinking habits of more than thirty thousand households, dividing drinkers into abstainers, moderate drinkers, and heavy drinkers (117). A moderate drinker was defined as one who drank from four to thirteen drinks per week. A heavy drinker averaged two or more drinks per day, or over fourteen drinks per week. A drink was defined as a 1½ ounce highball, a bottle of beer, or 4 ounces of wine.

Using these definitions, the researchers found that 13 percent of men and 3 percent of women were heavy drinkers. Heavy drinkers were concentrated in the highest income groups (families earning $50,000 and over) and among whites. There is some inconsistency about the social-class correlation as higher rates among blue-collar workers and urban blacks have been found in other studies, including an earlier nationwide survey of drinking practices by Cahalan, Cisin, and Crossley (11). Regarding drinkers in general, this survey also demonstrated that more drinkers live in cities and suburbs than in rural regions and small towns. To some extent religion determined whether a person was a drinker or a teetotaler: Almost all urban Jews and Episcopalians drank on occasion; fewer than half of rural Baptists drank. The drinking patterns revealed by this study were highly changeable. It was common for individuals to be heavy drinkers for long periods and then become moderate drinkers or teetotalers.

There have been no nationwide studies of prevalence rates of alcoholism in the United States. In part this is because of disagreement about the definition of alcoholism. Estimates of the extent of alcoholism range from 5 to 9 million Americans. These estimates are roughly equivalent to expectancy rates obtained in studies by Luxenberger (80) in Germany, Bleuler (5) in Switzerland, Sjögren (111) in Sweden, Fremming (30) in Denmark, and Slater (112) in England. In these countries the lifelong expectancy rate for alcoholism among men is about 3 to 5 percent; the rate for women ranges from 0.1 to 1 percent. Weissman and associates found similar rates in a late-1970s household survey in Connecticut (125).

In the United States, blacks in urban ghettos appear to have a particularly high rate of alcohol-related problems; whether rural blacks have comparably high rates of alcohol problems is unknown (83, 100). American Indians are also said to have high rates of alcoholism (23), but this does not apply to all tribes, and books have been written debunking the "firewater myth" (54, 73). Orientals, on the other hand, generally have low rates of alcoholism.

Alcoholism is a serious problem in France and Russia (106). Although Italy, like France, is a "vinocultural" country (where wine is a popular beverage), it is commonly asserted that Italians have a lower rate of alcoholism than the French (3, 78). The evidence for this is scant however. Estimates of alcoholism rates are usually based on cirrhosis rates and admissions to psychiatric hospitals for alcoholism. France has the highest cirrhosis rate in the world, but Italy also has a high rate, suggesting that alcoholism may be more common in Italy than is generally assumed. Ireland has a relatively low cirrhosis rate, despite its reputation for a high rate of alcoholism. According to de Lint and Schmidt (77), there is a positive correlation between per capita consumption of alcohol and cirrhosis rates. If cirrhosis rates are a reliable indicator of the extent of alcoholism in a particular country, then both cirrhosis and alcoholism are correlated with the total amount of alcohol consumed by the population of that country. Cirrhosis rates themselves are probably somewhat unreliable, and estimates of the prevalence of alcoholism in different countries tend to change from time to time as new information is obtained (97).

Alcohol problems are correlated with a history of school difficulty (85). High school dropouts and individuals with a record of frequent truancy and delinquency appear to have a particularly high risk of alcoholism.

No systematic studies have explored the relationship between occupation and alcoholism, but cirrhosis data suggest that individuals in certain occupations are more vulnerable to alcoholism than those doing other types of work. Waiters, bartenders, longshoremen, musicians, authors, and reporters have relatively high cirrhosis rates; accountants, mail carriers, and carpenters have relatively low rates (77).

Age is another demographic variable correlated with alcoholism. Urban blacks begin drinking at an earlier age than whites of comparable socioeconomic status (100). Women as a rule begin heavy drinking at a later age than men (37, 95).

Clinical Picture

Alcoholism is a behavioral disorder. The specific behavior that causes problems is the consumption of large quantities of alcohol on repeated occasions (Table 7.1). The motivation underlying this behavior is often obscure. When asked why they drink excessively, alcoholics occasionally attribute their drinking to a particular mood such as depression or anxiety or to situational problems. They sometimes describe an overpowering "need" to drink, variously described as a craving or compulsion. Just as often, however, the alcoholic is unable to give a plausible explanation of his excessive drinking (79).

Like other drug dependencies, alcoholism is accompanied by a preoccupation with obtaining the drug in quantities sufficient to produce intoxication over long periods. It is especially true early in the course of alcoholism that the patient may deny this preoccupation or attempt to rationalize his need by assertions that he drinks no more than his friends. As part of this denial or ratio-

Table 7.1 Diagnostic criteria for Alcohol Dependence (DSM-III-R)

Three or more manifestations required for diagnosis:

(1) alcohol often taken in larger amounts or over a longer period than the person intended

(2) persistent desire or one or more unsuccessful efforts to cut down or control alcohol use

(3) a great deal of time spent in activities necessary to get alcohol, taking the substance, or recovering from its effects

(4) frequent intoxication or withdrawal symptoms when expected to fulfill major role obligations at work, school, or home, (e.g., does not go to work because hung over, goes to school or work intoxicated, or intoxicated while taking care of his or her children), or when alcohol use is physically hazardous (e.g., drives when intoxicated)

(5) important social, occupational, or recreational activities given up or reduced because of alcohol

(6) continued alcohol use despite knowledge of having a persistent or recurrent social, psychological, or physical problem that is caused or exacerbated by the use of the substance

(7) marked tolerance: need for markedly increased amounts of the substance (i.e., at least a 50% increase) in order to achieve intoxication or desired effect, or markedly diminished effect with continued use of the same amount

(8) characteristic withdrawal symptoms

(9) alcohol often taken to relieve or avoid withdrawal symptoms

Criteria for determining degree of severity:

Mild: Few, if any, symptoms in excess of those required to make the diagnosis and the symptoms result in no more than mild impairment in occupational functioning or in usual social activities or relationships with other.

Moderate: Symptoms or functional impairment between "mild" and "severe."

Severe: Many symptoms in excess of those required to make the diagnosis, and the symptoms markedly interfere with occupational functioning or with usual social activities or relationships with others.

In partial remission: During the past six months, some use of alcohol and some symptoms of dependence.

In full remission: During the past six months, either no use of alcohol or use of alcohol and no symptoms of dependence.

nalization, alcoholics tend to spend their time with other heavy drinkers.

An alcoholic named David writes (47):

My need was easy to hide from myself and others (maybe I'm kidding myself about the others). I only associated with people who drank. I married a woman who drank. There were always reasons to drink. I was low, tense, tired, mad, happy. I probably drank as often because I was happy as for any other reason. And occasions for drinking—when drinking was appropriate, expected—were endless, Football games, fishing trips, parties, holidays, birthdays, Christmas, or merely Saturday night. Drinking became interwoven with everything pleasurable—food, sex, social life. When I stopped drinking, these things, for a time, lost all interest for me, they were so tied to drinking.

As alcoholism progresses and problems from drinking become more serious, alcoholics may drink alone, sneak drinks, hide the bottle, and take other measures to conceal the seriousness of their condition. This is almost always accompanied by feelings of guilt and remorse, which in turn may produce more drinking, temporarily relieving the feelings. Remorse may be particularly intense in the morning, when the patient has not had a drink for a number of hours, and this may provoke morning drinking (63):

For years [David writes], I drank and had very little hangover, but now the hangovers were gruesome. I felt physically bad—headachy, nauseous, weak, but the mental part was the hardest. I loathed myself. I was waking early and thinking what a mess I was, how I had hurt so many others and myself. The words "guilty" and "depression" sound superficial in trying to describe how I felt. The loathing was almost physical—a dead weight that could be lifted in only one way, and that was by having a drink, so I drank, morning after morning. After two or three, my hands were steady, I could hold some breakfast down, and the guilt was gone, or almost.

Prolonged drinking, even if initiated to relieve guilt and anxiety, commonly produces anxiety and depression (21). The full range of symptoms associated with depression and anxiety disorders—including terminal insomnia, low mood, irritability, and anxiety attacks with chest pain, palpitations, and dyspnea—often appear. Alcohol temporarily relieves these symptoms, resulting in a vicious cycle of drinking-depression-drinking, which may ultimately result in a classical withdrawal syndrome. Often the patient makes a valiant effort to stop drinking and may succeed for a period of several days or weeks, only to fall off the wagon again:

At some point I was without wife, home, or job. I had nothing to do but drink. The drinking was now steady, days on end. I lost appetite and missed meals (besides money was short). I awoke at night, sweating and shaking, and had a drink. I awoke in the morning vomiting and had a drink. It couldn't last. My ex-wife found me in my apartment shaking and seeing things, and got me in the hospital. I dried out, left, and went back to drinking. I was hopitalized again, and this time stayed dry for six months. I was nervous and couldn't sleep, but got some of my confidence back and found a part-time job. Then my ex-boss offered my job back and I celebrated by having a drink. The next night I had two drinks. In a month I was drinking as much as ever and again unemployed.

Repeated experiences like this easily lead to feelings of despair and hopelessness. By the time patients consult a physician, they have often reached rock bottom. Their situation seems hopeless and, after years of heavy drinking, their problems have become so numerous that they feel nothing can be done about them. At this point, they may be ready to acknowledge their alcoholism, but feel powerless to stop drinking. But many do stop—permanently—as will be discussed later.

Alcohol is one of the few psychoactive drugs that produce, on occasion, classical amnesia. Nonalcoholics, when drinking, also experience this amnesia (blackouts), but much less often, as a rule, than do alcoholics (36). These episodes of amnesia are particularly distressful to alcoholics because they may fear that they have unknowingly harmed someone or behaved imprudently while intoxicated (37):

A thirty-nine-year-old salesman awoke in a strange hotel room. He had a mild hangover but otherwise felt normal. His clothes were hanging in the closet; he was clean-shaven. He dressed and went down to the lobby. He learned from the clerk that he was in Las Vagas and that he had checked in two days previously. It had been obvious that he had been drinking, the clerk said, but he hadn't seemed very drunk. The date was Saturday the fourteenth. His last recollection was of sitting in a St. Louis bar on Monday the ninth. He had been drinking all day and was drunk, but could remember everything perfectly until about three P.M., when "like a curtain dropping," his memory went blank. It remained blank for approximately five days. Three years later, those five days were still a blank. He was so frightened by the experience that he abstained from alcohol for two years.

Studies of blackouts (37, 113) indicate that the amnesia is anterograde. During a blackout, individuals have relatively intact remote and immediate memory but they experience a specific short-term memory deficit in which they are unable to recall events that happened five or ten minutes before. Because their other intellectual faculties are well preserved, they can perform complicated acts and appear normal to the casual observer. Present evidence suggests that alcoholic blackouts represent impaired consolidation of new information rather than repression motivated by a desire to forget events that happened while drinking (37).

Sometimes, however, a curious thing happens: The drinker recalls things that happened during a previous drinking period that, when sober, he had forgotten. For example, alcoholics often report hiding money or alcohol when drinking, forgetting it when sober, and having their memory return when drinking again (37):

A forty-seven-year-old housewife often wrote letters when she was drinking. Sometimes she would jot down notes for a letter and start writing it but not finish it. The next day, sober, she would be unable to decipher the notes. Then she would start drinking again, and after a few drinks the meaning of the notes would become clear and she would resume writing the letter. "It was like picking up the pencil where I had left off."

Anecdotal reports suggest that benzodiazepine medication combined with alcohol increases the likelihood of a blackout occurring with smaller amounts of alcohol than is usually required (90).

Benzodiazepines taken orally produce a mild impairment of short-term memory (71) but are not associated with blackouts of the type produced by alcohol (unless given intravenously).

By the time alcoholics consult a physician, they have often developed medical and social complications from drinking (see Complications).

Identifying the Alcoholic

Before alcoholism can be treated, it first must be recognized. Physicians are in a particularly good position to identify a drinking problem early: They can do a physical examination and order laboratory tests. Here are some hints of a drinking problem:

1. Arcus senilis—a ringlike opacity of the cornea—occurs commonly with age, causes no visual disturbance, and is considered an innocent condition. The ring forms from fatty material in the blood. Alcohol increases fat in the blood and more alcoholics are reported to have the ring than others their age (27).

2. A red nose (acne rosacea) suggests the owner has a weakness for alcoholic beverages. Often, however, people with red noses are teetotalers or even rabid prohibitionists, and they resent the insinuation.

3. Red palms (palmer erythema) are also suggestive but not diagnostic of alcoholism.

4. Cigarette burns between the index and middle fingers or on the chest and contusions and bruises should raise suspicions of alcoholic stupor.

5. Painless enlargement of the liver may suggest a larger alcohol intake than the liver can cope with. Severe, constant upper abdominal pain and tenderness radiating to the back indicates pancreatic inflammation, and alcohol sometimes is the cause.

6. Reduced sensation and weakness in the feet and legs may occur from excessive drinking.

7. Laboratory tests provide other clues. More than half of alcoholics have increased amounts of gamma-glutamyl transpeptidase (GGT) in their blood, which is unusual in nonalcoholics. After

GGT, elevations in the following tests are most often associated with heavy drinking: mean corpuscular volume, uric acid, triglycerides, aspartate aminotransferase, and urea (126).

Another approach is to use a wide range of commonly available blood chemistry tests and subject them to quadratic discriminant analysis (105). Each value may be in the normal range, but *in toto* they produce a distinctive "fingerprint" that is highly specific for detecting recent heavy drinking. Whether it is a useful test for alcoholism per se is not known.

In their search for signs of alcohol abuse, physicians sometimes slip into a moralistic attitude that alienates the patient. For personal reasons, physicians may believe any drinking is wrong, but they still should be aware of new findings that suggest moderate drinking may actually contribute to longevity (2, 4, 69). Turner and associates (116) in their review of the "beneficial" effects of alcohol surveyed a number of well-designed studies that indicated the risk of coronary heart disease is lower in persons who use alcohol moderately than in abstainers. Increased levels of high-density lipoproteins (HDL)—a negative risk factor for myocardial infarction—are associated with moderate and even (in some studies) heavy alcohol use. Moderate drinking is defined by Turner et al. as an intake that exceeds neither 0.8 g/kg of ethanol per day (up to a limit of 80 g) nor an average of 0.7 g/kg for three successive days.

None of this should be interpreted as encouragement of immoderate use of alcohol. For alcoholics, no level of intake is considered beneficial. Moreover, uncertainties exist concerning both the effects of moderate drinking on the unborn child and the relationship of moderate drinking to automobile accidents.

Natural History

The natural history of alcoholism seems to be somewhat different in men and women (95). In men the onset is usually in the late teens or twenties, the course is insidious, and the alcoholic may not be fully aware of his dependency on alcohol until his thirties.

The first hospitalization usually occurs in the late thirties or forties (24).

In men symptoms of alcoholism rarely occur for the first time after age forty-five (47). If they do occur, the physician should be alerted to the possibility of primary affective disorder or brain disease.

Alcoholism has a higher "spontaneous" remission rate than is often recognized. The incidence of first admissions to psychiatric hospitals for alcoholism drops markedly in the sixth and seventh decades, as do first arrests for alcohol-related offenses. Although the mortality rate among alcoholics is perhaps two to three times that of nonalcoholics, this is probably insufficient to account for the apparent decrease in problem drinking in middle and late middle life (24).

Women alcoholics have been studied less extensively than men alcoholics, but the evidence suggests that the course of the disorder is more variable in women. The onset often occurs later and spontaneous remission apparently is less frequent (37, 95). Women alcoholics are also more likely to have a history of primary affective disorder (107).

On the basis of questionnaire data obtained from alcoholics, Jellinek promulgated the view that manifestations of alcoholism follow a natural chronological order, with blackouts being one of the early "prodromal" symptoms of the illness (63). Later studies (36) have challenged this view, and it is now believed that problems from drinking may occur in various sequences and that blackouts have no special significance as a sign of incipient alcoholism. Frequently, after years of heavy problem-free drinking, a person may experience a large number of problems in a brief period. Table 7.2 shows the mean age of onset of alcohol problems in an unselected series of hospitalized men alcoholics.

Patterns of drinking also are variable, and it is a mistake to associate one particular pattern exclusively with alcoholism. Jellinek (62) divided alcoholics into various "species," depending on their pattern of drinking. One species, the so-called gamma alcoholic, is common in America and conforms to the stereotype of

Table 7.2 Onset of alcoholic manifestations among all subjects (N = 100)

Manifestation	Present (%)	Mean Age at Onset	Percentage of Subjects Reporting Manifestation (by age at first occurrence*)										
			<20	20–24	25–29	30–34	35–39	40–44	45–49	50–54	55–59	60–64	65–70
1. Frequent drunks	98	27	18	17	21	18	13	7	3		1	1	
2. Weekend drunks	82	28	11	18	21	32	6	8	2		1	1	
3. Morning drinking	84	31	2	17	15	25	17	9	11	1	1		
4. Benders	76	31	7	14	18	21	13	9	12	5	1		
5. Neglecting meals	86	32	2	11	14	29	15	10	10	6	1		1
6. "Shakes"	88	33	1	11	15	24	17	9	16	3	2		1
7. Job loss from drinking	69	34	3	7	19	19	26	7	9	6	3	1	
8. Separation or divorce from drinking	44	34		16	7	23	14	20	14	7			
9. Blackouts	64	35	2	6	16	28	25	8	9	3	3		
10. Joined AA	39	36		8	8	28	20	15	8	10	3		
11. Hospitalization for drinking	100	37	1	3	14	18	20	13	14	9	6	1	
12. Delirium tremens	45	38		2	11	11	40	16	7	11	2		1

* Figures rounded off to nearest whole number for ease of perusal. Sums may therefore not equal 100 percent.

Goodwin et al. (36).

the Alcoholics Anonymous member. Gamma alcoholics have problems with "control": Once they begin drinking, they are unable
to stop until poor health or depleted financial resources prevent
them from continuing. Once the bender is terminated, however,
the person is able to abstain from alcohol for varying lengths of
time. Jellinek contrasted the gamma alcoholic with a species of
alcoholic common in France. He has "control" but is "unable to
abstain"; he *must* drink a given quantity of alcohol every day,
although he has no compulsion to exceed this amount. He may
not recognize that he has an alcohol problem until, for reasons
beyond his control, he has to stop drinking, whereupon he experiences withdrawal symptoms.

Although these pure types of alcoholism do exist, many individuals who do not conform to the stereotypes still have serious
drinking problems. Among American alcoholics, one drink does
not invariably lead to a binge; a person may drink moderately for
a long time before his or her drinking begins to interfere with
health or social functioning (118).

This diversity in drinking patterns explains the current emphasis
on *problems* rather than a single set of symptoms as the basis for
diagnosing alcoholism.

Complications

Because alcoholism is defined by the problems it creates, symptoms and complications inevitably overlap. For present purposes,
we will consider social and medical complications separately.

Alcoholics have a high rate of marital separation and divorce
(102). They often have job troubles, including frequent absen-
teeism and job loss. They also have a high frequency of accidents—in the home, on the job, and while driving automobiles.
About 40 percent of highway fatalities in the United States involve a driver who has been drinking (110). Nearly half of convicted felons are alcoholic (38) and about half of police activities
in large cities are associated with alcohol-related offenses.

Medical complications fall into three categories: (a) acute effects of heavy drinking, (b) chronic effects of heavy drinking, and (c) withdrawal effects.

Consumption of very large amounts of alcohol can lead directly to death by depressing the respiratory center in the medulla. Acute hemorrhagic pancreatitis occasionally occurs from a single heavy drinking episode.

Nearly every organ system can be affected, directly or indirectly, by chronic, heavy use of alcohol. Gastritis and diarrhea are common reversible effects. Gastric ulcer may also occur, although the evidence that alcohol directly produces ulceration is equivocal (124).

The most serious effect of alcohol on the gastrointestinal tract is liver damage. After many years of study, however, it is still not clear whether alcohol alone has a direct toxic effect on the liver (33). At present, it appears that cirrhosis results from the combined effect of alcohol and diet plus other factors, possibly including heredity. Human and animal studies indicate that a single large dose of alcohol combined with a diet rich in fat produces a fatty liver (75, 124). Conversely, alcohol together with fasting can result in a fatty liver. The connection between fatty liver and cirrhosis, howover, is unclear. The fatty changes in the liver after acute alcohol intoxication are reversible. Most patients with Laennec's cirrhosis in Western countries are excessive drinkers. Most severe alcoholics, however, do not develop cirrhosis (probably less than 10 percent).

Studies show that force-feeding baboons with large amounts of alcohol over long periods results in cirrhosis (76). However, not all baboons develop cirrhosis, and these studies leave unresolved the question of the direct role of alcohol in producing cirrhosis. One complication in interpreting the finding is that large intake of alcohol also produces a malabsorption syndrome. Therefore, although the baboons had adequate diets, it is likely that important food constituents, including vitamins, were not fully absorbed during the drinking periods.

Alcoholism is associated with pathology of the nervous system, principally owing to vitamin deficiencies and not to a direct toxic effect of alcohol. Peripheral neuropathy, the most common neurological complication, apparently results from multiple vitamin B deficiencies (119). It is usually reversible with adequate nutrition. Retrobulbar neuropathy may lead to amblyopia (sometimes called tobacco-alcohol amblyopia), which is also usually reversible with vitamin therapy.

Other neurological complications include anterior lobe cerebellar degenerative disease (102) and the Wernicke-Korsakoff syndrome (120). The latter results from thiamine deficiency (38). The acute Wernicke stage consists of ocular disturbances (nystagmus or sixth nerve palsy), ataxia, and confusion. It usually clears in a few days but may progress to a chronic brain syndrome (Korsakoff psychosis). Short-term memory loss (anterograde amnesia) is the most characteristic feature of Korsakoff's psychosis. "Confabulation" (narration of fanciful tales) may also occur. The Wernicke-Korsakoff syndrome is associated with necrotic lesions of the mammillary bodies, thalamus, and other brain stem areas. Thiamine corrects early Wernicke signs rapidly and *may* prevent development of an irreversible Korsakoff dementia. Once the dementia is established, thiamine does not usually help.

A 48-year-old divorced housepainter has been admitted to the hospital with a history of 30 years of heavy drinking. He has had two previous admissions for detoxification but the family states that he has not had a drink in several weeks and shows no sign of alcohol withdrawal. He looks malnourished, however, and on examination is found to be ataxic and have a bilateral sixth cranial nerve palsy. He appears confused and mistakes one of his physicians for a dead uncle.

Within a week the patient walks normally and there is no sign of a sixth nerve palsy. He seems less confused and can now find his way to the bathroom without direction. He remembers the names and birthdays of his siblings but has difficulty naming the past five presidents.

More strikingly, he has great difficulty in retaining information for longer than a few minutes. He can repeat back a list of numbers immediately after he has heard them, but a few minutes later does not recall being asked to perform the task. Shown three objects (keys, comb, ring), he cannot recall them three minutes later. He does not seem worried about this. Asked if he can recall the name of his doctor, he replies, "Certainly," and proceeds to call the doctor "Dr. Masters" (not his name) whom he claims he first met in the Korean War. He tells a long untrue story about how he and "Dr. Masters" served as fellow soldiers in the Korean War.

The patient is calm, alert, friendly. One could be with him for a short period and not realize he has a severe memory impairment, in view of his intact immediate memory and spotty but sometimes impressive remote memory. His amnesia, in short, is largely anterograde. Although treated with high doses of thiamine, the short-term memory deficit persists and appears to be irreversible (48).

Whether excessive use of alcohol produces cortical atrophy has been debated for many years (20). Computerized tomography (CT) scans of alcoholics have been contradictory, some showing cerebral atrophy (26, 61), some showing none (58), and one showing *reversible* atrophy (12). In a 1987 study, computer technology was used to count brain cells in alcoholics who came to autopsy; alcoholics had fewer neurons in the frontal cortex than did nonalcoholics (55). If confirmed by further studies, this would support the long-held but poorly documented view that excessive use of alcohol produces cortical atrophy of the frontal lobes. The extensive psychometric literature on intellectual impairment in alcoholics is also contradictory (43). Most studies performed soon after drinking bouts show intellectual deficits, but the deficits vary from study to study. The most consistent results have involved use of the categories test of the Halstead Battery, indicating that alcoholics have difficulty in "conceptual shifting." Many of the deficits found in alcoholics undergoing detoxification are reversible. No studies have reported a decline in alcoholics' IQs.

Other medical complications of alcoholism include cardiomyopathy, thrombocytopenia and anemia, and myopathy. Two recent studies showed an increase in breast cancer in women who were light or moderate drinkers (56, 127). Having not previously been reported, the association awaits further study for confirmation. It has also been reported that alcoholics have an increased risk of stroke (34) This could be related to alcohol-induced hypertension, although the relationship between alcohol and sustained hypertension remains controversial (82). The effect of alcohol on stroke was lost when adjusted for smoking in another study (53).

A possible teratogenic effect of alcohol has been suspected for centuries, but it was not until the work of Lemoine in 1968 (74) and the independent observations of Jones and Smith in 1973 (64) that a distinct dysmorphic condition associated with maternal alcoholism during pregnancy was described in the medical literature. The abnormalities most typically associated with alcohol teratogenicity can be grouped into four categories: central nervous system dysfunction, birth deficiencies, a characteristic cluster of facial abnormalities, and variable major and minor malformations (17). Animal studies have now demonstrated specific teratogenic properties of ethyl alcohol in a variety of species, many of the abnormalities being similar to those described in man (15, 98).

Despite this growing body of evidence, the so-called fetal alcohol syndrome remains a subject of controversy. Large-scale longitudinal studies are needed to definitely establish the teratogenicity of alcohol in humans. This is because alcoholic women are often malnourished, take other drugs, smoke heavily, and have life styles in general that differ from those of nonalcoholic women. Most clinicians considering present evidence believe women should be cautioned against drinking excessively during pregnancy and perhaps should avoid alcohol altogether as "safe" levels of alcohol have not been ascertained (27, 37).

The term "alcohol withdrawal syndrome" is preferable to "delirium tremens" (DTs). The latter refers to a specific manifestation of the syndrome. The most common withdrawal symptom is tremulousness, which usually occurs only a few hours after cessation of drinking and may even begin while the person is still drinking ("relative abstinence"). Transitory hallucinations also may occur. If so, they usually begin twelve to twenty-four hours after drinking stops (119).

After a week of heavy drinking and little food, a 30-year-old newspaper reporter tried to drink a morning cup of coffee and found that his hands were shaking so violently he could not get the cup to his mouth. He managed to pour some whisky into a glass and drank as much as he could. His hands became less shaky but now he was nauseated and began having "dry heaves." He tried repeatedly to drink but could not keep the alcohol down. He felt ill and intensely anxious and decided to call a doctor friend. The doctor recommended hospitalization.

On admission, the patient had a marked resting and exertional tremor of the hands, and his tongue and eyelids were tremulous. He also had feelings of "internal" tremulousness. Lying in the hospital bed, he found the noises outside his window unbearably loud and began seeing "visions" of animals and, on one occasion, a dead relative. He was terrified and called a nurse, who gave him a tranquilizer. He became quieter and his tremor was less pronounced. At all times he realized that the visual phenomena were "imaginary." He always knew where he was and was oriented otherwise. After a few days the tremor disappeared and he no longer hallucinated. He still had trouble sleeping but otherwise felt back to normal and vowed never to drink again (48).

Grand mal convulsions (rum fits) occur occasionally, sometimes as long as two or three days after drinking stops. As a rule, alcoholics experiencing convulsions do not have epilepsy; they have normal electroencephalograms (EEGs) when not drinking and experience convulsions only during withdrawal (84).

Delirium tremens is infrequent and when it does occur is often associated with an intervening medical illness. For a diagnosis of delirium tremens, gross memory disturbance should be present in addition to other withdrawal symptoms such as agitation and vivid hallucinations. Classically, delirium tremens begins two or three days after drinking stops and subsides within one to five days (119). One must always suspect intercurrent medical illness when delirium occurs during withdrawal. The physician should be particularly alert to hepatic decompensation, pneumonia, subdural hematoma, pancreatis, and fractures.

So-called chronic alcoholic hallucinosis refers to the persistence of hallucinations, usually auditory, for long periods after other abstinence symptoms subside and after the patient has stopped heavy drinking (121) This occurs rarely; after seventy-five years of debate concerning the etiology, it has not been resolved whether drinking actually produces the condition.

Suicide is an important complication of alcoholism. About one-quarter of suicides are alcoholic, predominantly white men over age thirty-five. Apparently alcoholics (unlike patients with primary affective disorder) are especially likely to commit suicide after loss of a wife, close relative, or other serious interpersonal disruption (39).

Family Studies

Every family study of alcoholism regardless of country of origin has shown much higher rates of alcoholism among the relatives of alcoholics than in the general population. According to a number of studies, about 25 percent of fathers and brothers of alcoholics are themselves alcoholics (19, 47).

Not everything that runs in families is inherited, however. Speaking French, for example, runs in families and is not inherited. How does one separate nature from nurture in familial illnesses?

Two methods are used in psychiatric research. One is to compare identical and fraternal twins. If identical twins are more often concordant for alcoholism than fraternal twins, this would point to a genetic influence. Another way is to study adopted persons who have alcoholism in the biological family but not in the family of upbringing. Both approaches have been applied to alcoholism. Here, briefly, are the results.

Twin Studies. A Swedish investigator (66) found that identical twins were significantly more concordant for alcoholism than fraternal twins, the more severe the alcoholism the greater the difference. Among younger persons in a Finnish study (93), there was a difference between identical and fraternal twins with regard to alcohol problems, but there was no difference in the total sample. A third study, in England (91), showed no difference between identical and fraternal twins. The most recent study (59) involved analyzing a large number of Veterans Administration records. The results supported a genetic factor; identical twins were more often concordant for alcoholism than fraternal twins. In summary, two twin studies produced results consistent with a genetic influence, one did not, and a fourth was equivocal.

Adoption Studies. The first adoption study of alcoholism was published in 1944 (101). It showed no difference in drinking behavior between children of alcoholics and children of nonalcoholics, both in their early twenties. The sample size was small and no criteria were presented for the diagnosis of alcoholism. It was not clear whether the biological parents were alcoholic by today's definition.

A small flurry of adoption studies began in the early 1970s, ushered in by a study comparing adult half-siblings of alcoholics with adult full-siblings (108). The assumption behind the study was that if genetic factors were important, full-siblings would more often be alcoholic than half-siblings. This did not prove to be the case. However, the study did find that half-siblings, predictably, were from broken families, and this afforded the opportunity to compare the incidence of alcoholism in offspring with alcoholism in biological parents and in parents of upbringing. Having a biological father who was alcoholic was highly correlated with alcoholism in the sons, but there was no correlation with alcoholism in the surrogate fathers. The study, in short, failed to prove its central hypothesis but ended up as a kind of adoption study suggesting that biological factors were more important than environmental factors in producing alcohol problems.

In the decade after the hallf-sibling study, the results from three separate adoption studies in Denmark, Sweden, and the United States (Iowa) were published (7, 9, 40). The studies produced remarkably similar findings: (1) sons of alcoholics were three or four times more likely to be alcoholic than were sons of nonalcoholics, whether raised by their alcoholic biological parents or by nonalcoholic adoptive parents; (2) sons of alcoholics were no more susceptible to nonalcoholic adult psychiatric disturbances (e.g., depression, sociopathy) than were sons of nonalcoholics where both groups were raised by nonalcoholic adoptive parents. The Iowa study (9, 10) did find a higher rate of childhood conduct disorder in the male offspring of alcoholics, an association later supported by several studies that have documented a history of hyperactive syndrome or conduct disorder in the childhood of alcoholics (42, 114)

The results regarding female alcoholism were more equivocal. In the Danish study (45) 4 percent of the daughters of alcoholics were alcoholics, but so were 4 percent of the control women, introducing the possibility that adoption could contribute to alcoholism. A low rate of alcoholism in the adopted-out daughters of alcoholics was found in the Swedish study (7, 8), although subsequent analysis revealed a correspondence between alcoholism in the biological mothers and alcoholism in the adopted-out daughters (18).

The analysis of this study has undergone a series of statistical refinements by Cloninger et al. (18) in the United States. These authors proposed that alcoholics could be divided into two types. One involves only men with severely alcoholic fathers where environmental factors were relatively unimportant. The other involves both sexes with a milder form of parental alcoholism in which environmental factors (e.g., low social class) were important in determining whether alcoholism occurred. The hypothesis underwent some subsequent modification, but remains interesting and testable.

Thus, studies condicted independently in three countries suggest that a susceptibility to alcoholism is transmitted from parent to child irrespective of whether the child is exposed to the alcoholic parent. This represents the strongest evidence to date for a genetic influence in alcoholism—or at least in *some* cases of alcoholism.

The Concept of Familial Alcoholism. Toward the end of the 1970s, investigators began taking note of something rather obvious but often overlooked in recent years: Many alcoholics do not have alcoholism in the family. In most studies, about half of hospitalized alcoholics give a family history of alcoholism and about half do not. In the past few years, a number of investigators have compared the two groups. Two consistent findings have emerged (52): (1) familial alcoholics (as they are called) show the first signs of dependence at a younger age than do nonfamilial alcoholics; (2) familial alcoholics have a more severe form of dependence with a more rapid, fulminating course.

Other differences are reported less consistently. Several studies (10, 52) indicate that familial alcoholics are less likely to have other psychiatric illnesses than nonfamilial alcoholics. In this sense, familial alcoholism is synonymous with "primary" alcoholism. An exception to the general rule that *familial* equals *primary* is the finding by two groups that familial alcoholics more often have antisocial personalities (57, 96). This is still controversial. There is more agreement that familial alcoholics often have a history of childhood hyperactivity and/or conduct disorder (114).

We now have sufficient grounds for tentatively viewing familial and nonfamilial alcoholism as separate conditions. One study even suggests that this variable may be a predictor of treatment outcome: Nonfamilial alcoholics more often appeared to return to "controlled" drinking after treatment than familial alcoholics (87). Obviously, more studies are needed before this should influence clinical management in any way. That family history should be included as a potentially important variable in future research on alcoholism, however, is clear.

Differential Diagnosis

Chronic excessive use of alcohol produces a wide range of psychiatric symptoms that, in various combinations, can mimic other psychiatric disorders. Therefore, while a person is drinking heavily and during the withdrawal period, it is difficult to determine whether he or she suffers from a psychiatric condition other than alcoholism.

The diagnosis of alcoholism itself is relatively easy. However, many alcoholics also use other drugs, and it may be difficult to determine which symptoms are produced by alcohol and which by barbiturates, amphetamines, and so on. If a patient has been drinking heavily and not eating, he may become hypoglycemic (29), and this condition may produce symptoms resembling those seen in withdrawal.

The two psychiatric conditions most commonly associated with alcoholism are primary affective disorder and sociopathy (50). Female alcoholics apparently suffer more often from primary affective disorder than do male alcoholics (107). The diagnosis of primary affective disorder usually can be made by past history or by observing the patient during long periods of abstinence. According to one study, about one third of patients with manic depressive illness drink more while depressed and another third drink less (13). Studies indicate that small amounts of alcohol administered to a depressed patient relieve depressive symptoms, but large amounts worsen depression (85).

Many sociopaths drink to excess, although how many would be considered "alcoholic" is uncertain. A follow-up study of convicted felons, about half of whom had alcohol problems, indicates that sociopathic drinkers have a higher "spontaneous" remission rate than do nonsociopathic alcoholics (36, 38). When sociopaths reduce their drinking, their criminal activities are correspondingly reduced.

Various personality disorders have been associated with alcoholism, particularly those in which "dependency" is a feature. The consensus at present is that alcoholism is not connected with a particular constellation of personality traits. Longitudinal studies help little in predicting what types of individuals are particularly susceptible to alcoholism (65, 86).

Clinical Management

The treatment of alcoholism and the management of alcohol withdrawal symptoms present separate problems.

In the absence of serious medical complications, the alcohol withdrawal syndrome is usually transient and self-limited; the patient recovers within several days regardless of treatment (119). Insomnia and irritability may persist for longer periods (33).

Treatment for withdrawal is symptomatic and prophylactic. Relief of agitation and tremulousness can be achieved with a variety of drugs, including barbiturates, paraldehyde, chloral hydrate, the phenothiazines, and the benzodiazepines. Currently, the benzodiazepines (e.g., chlordiazepoxide) are widely considered the drugs of choice for withdrawal. They have little, if any, synergistic action with alcohol and, compared with barbiturates and paraldehyde, relatively little abuse potential. They can be administered parenterally to intoxicated patients without apparent risk and continued orally during the withdrawal period. There is some evidence that mortality is increased when the phenothiazines are used, reportedly from hypotension or hepatic encephalopathy.

Administration of large doses of vitamins—particularly the B vitamins—is obligatory, given the role of these vitamins in preventing peripheral neuropathy and the Wernicke-Korsakoff syndrome (38). The B vitamins are water-soluble and there is no apparent danger in administering them in large doses.

Unless the patient is dehydrated because of vomiting or diarrhea, there is no reason to administer fluids parenterally. Contrary to common belief, alcoholics usually are not dehydrated; actually, they may be overhydrated from consumption of large volumes of fluid (92). During the early stages of withdrawal, hyperventilation may cause respiratory alkalosis and this, together with hypomagnesemia, has been reported to produce withdrawal seizures (87). If the individual has a history of withdrawal seizures, diphenylhydantoin (Dilantin) may be prescribed, although there is no evidence that it prevents withdrawal seizures.

If patients develop delirium, they should be considered dangerous to themselves and others, and protective measures taken. Ordinarily, tranquilizers will calm the patient sufficiently to control agitation, and restraints will not be necessary. Administration of intravenous barbiturates may be necessary for severe agitation. Most important, if delirium occurs, further exploration should be conducted to rule out serious medical illness missed in the original examination. When a patient is delirious, an attendant should always be present. It is sometimes helpful to have a friend or relative present.

The treatment of alcoholism should not begin until withdrawal symptoms subside. Treatment has two goals: (a) sobriety and (b) amelioration of psychiatric conditions associated with alcoholism. A small minority of alcoholics are eventually able to drink in moderation, but for several months after a heavy drinking bout total abstinence is desirable for two reasons. First, the physician must follow the patient, sober, for a considerable period to diagnose a coexistent psychiatric problem. Second, it is important for the patient to learn that he or she can cope with ordinary life problems without alcohol. Most relapses occur within six months of discharge from the hospital; they become less and less frequent after that (35).

For many patients, disulfiram (Antabuse) is helpful in maintaining abstinence. By inhibiting aldehyde dehydrogenase, the drug leads to an accumulation of acetaldehyde if alcohol is consumed. Acetaldehyde is highly toxic and produces nausea and hypotension. The latter condition in turn produces shock and may be fatal. In recent years, however, Antabuse has been prescribed in a lower dosage (250 mg) than was employed previously and no deaths from its use have been reported for a number of years. One study indicates the dose is irrelevant; the deterrent effect is psychological and not dose dependent (31).

Discontinuation of Antabuse after administration for several days or weeks still deters drinking for a three- to five-day period because the drug requires that long to be excreted. Thus, it may be useful to give patients Antabuse during office visits at three- to four-day intervals early in the treatment program.

Until recent years it was recommended that patients be given Antabuse for several days and challenged with alcohol to demonstrate the unpleasant effects that follow. This procedure was not always satisfactory because some patients showed no adverse effects after considerable amounts of alcohol were consumed and other patients became very ill after drinking small amounts of alcohol. At present, the alcohol challenge test is considered optional. The principal disadvantage of Antabuse is not that patients drink while taking the drug but that they stop taking the drug after a brief period. This, again, is a good reason to give the drug on frequent office visits during the early crucial period of treatment.

In recent years, a wide variety of procedures, both psychological and somatic, have been tried in the treatment of alcoholism. None has proven definitely superior to others (6). There is no evidence that intensive psychotherapy helps most alcoholics. Nor are tranquilizers or antidepressants usually effective in maintaining abstinence or controlled drinking (22, 49). Aversive conditioning techniques have been tried, with such agents as apomorphine and emetine to produce vomiting (123), succinylcholine to produce apnea (16), and electrical stimulation to produce pain (60). The controlled trials required to show that these procedures are effective have not been conducted, but a high rate of success has been reported for the apomorphine treatment in well-motivated patients (47). Although we do not know how many alcoholics benefit from participation in Alcoholics Anonymous, most clinicians agree that alcoholics should be encouraged to attend their meetings on a trial basis.

In three double-blind studies, lithium carbonate was found superior to placebo in reducing drinking in depressed alcoholics (28, 70, 88). There was a high drop-out rate in all three studies. A fourth study failed to confirm the efficacy of lithium for alcoholism (97).

In conclusion, it should be emphasized that relapses are characteristic of alcoholism and that physicians treating alcoholics should avoid anger or excessive pessimism when such relapses occur. Alcoholics see nonpsychiatric physicians as often as they see psychiatrists (probably more often), and there is evidence that general practitioners and internists are sometimes more helpful (32). This may be particularly true when the therapeutic approach is warm but authoritarian, with little stress on "insight" or "understanding." Since the cause of alcoholism is unknown, "understanding," in fact, means acceptance of a particular theory. That may provide temporary comfort but probably rarely provides lasting benefit.

References

1. Abel, E. L., and Sokol, R. J. Incidence of fetal alcohol syndrome and economic impact of FAS-related anomalies. Drug Alcohol Depend. 19: 51–70, 1987.

2. Anderson, K. M., Casteeli, W. P., and Levy, D. Cholesterol and mortality: 30 years of follow-up from the Framingham study. JAMA 257: 2176–2180, 1987.

3. Bales, R. F. Cultural differences in rates of alcoholism. Q. J. Stud. Alcohol 5:480, 1946.

4. Baum-Baucker, C. The health benefits of moderate alcohol consumption: a review of the literature. Drug and Alcohol Dependency 15:207–227, 1985.

5. Bleuler, M. Psychotische Belastung von körperlich Kranken. Z. Ges. Neurol. Psychiat. 142:780, 1932.

6. Blum, E. M., and Blum, R. H. Alcoholism. San Francisco: Jossey-Bass, 1969.

7. Bohman, M. Some genetic aspects of alcoholism and criminality: a population of adoptees. Arch. Gen. Psychiat. 35:269–276, 1978.

8. Bohman, M., Sigvardsson, S., and Cloninger, R. Maternal inheritance of alcohol abuse: cross-fostering analysis of adopted women. Arch. Gen. Psychiat. 38:965–969, 1981.

9. Cadoret, R. J. and Gath, A. Inheritance of alcoholism in adoptees. Brit. J. Psychiat. 132:252–258, 1978.

10. Cadoret, R. J., Cain, C. A., and Grove, W. M. Development of alcoholism in adoptees raised apart from alcoholic biologic relatives. Arch. Gen. Psychiat. 37:561–563, 1979.

11. Cahalan, D., Cisin, I. H., and Crossley, H. M. American Drinking Practices: A National Survey of Behavior and Attitudes, Monogr. no. 6. New Brunswick, N.J.: Rutgers Univ. Center of Alcohol Studies, 1969.

12. Carlen, P. L., Wortzman, G., Holgate, R. C., Wilkinson, D. A., and Rankin, J. G. Reversible cerebral atrophy in recently abstinent chronic alcoholics measured by computed tomography scans. Science 200:1076–1078, 1978.

13. Cassidy, W. L., Flanagan, N. B., Spellman, M., and Cohen, M. E. Clinical observations in manic-depressive disease: A quantitative study of 100 manic depressive patients in 50 medically sick controls. JAMA 164:1535–1546, 1953.

14. Chaucer, G. The Canterbury Tales: The Pardoner's Tale. In The Student's Chaucer, Skeats, W. W. (ed.). New York: Oxford Univ. Press, 1900.

15. Chernoff, G. F. The fetal alcohol syndrome in mice: an animal model. Teratology. 15:223–230, 1977.

16. Clancy, J., Vanderhoof, E., and Campbell, P. Evaluation of an aversive technique as a treatment for alcoholism. Q. J. Stud. Alcohol 28:476–485, 1967.

17. Clarren, S. K., and Smith, D. W.   The fetal alcohol syndrome. N. Engl. J. Med. 298:1063–1067, 1978.

18. Cloninger, C. R., Bohman, M., and Sigvardsson, S.   Inheritance of alcohol abuse: cross-fostering analysis of adopted men. Arch. Gen. Psychiat. 36:861–868, 1981.

19. Cotton, N. S.   The familial incidence of alcoholism: a review. J. Stud. Alcohol. 40:89–116, 1979.

20. Courville, C.   The effects of alcohol on the nervous system of man. San Lucas, Los Angeles, 1955. (Privately printed.)

21. Davis, D.   Mood changes in alcoholic subjects with programmed and free-choice experimental drinking. In *Recent Advances in Studies of Alcoholism*. Washington, D.C.: U.S. GPO, 1971.

22. Ditman, K. S.   Review and evaluation of current drug therapies in alcoholism. Int. J. Psych. 3:248–258, 1967.

23. Dozier, E. P.   Problem drinking among American Indians: the role of socio-cultural deprivation. Q. J. Stud. Alcohol 27:72–87, 1966.

24. Drew, L. R. H.   Alcoholism as a self-limiting disease. Q. J. Stud. Alcohol 29:956–967, 1968.

25. Edwards, G., Gross, M. M., Keller, M., Moser, J., and Room, R. (eds.).   Alcohol-related disabilities. WHO Offset Publ., no. 32 Geneva: WHO, 1977.

26. Epstein, P. S., Pisani, V. D., and Fawcett, J. A.   Alcoholism and cerebral atrophy. Alcoholism: Clin. Exp. Res. 1:61–65, 1977.

27. Ewing, J. A., and Rouse, B. A.   Corneal arcus as a sign of possible alcoholism. Alcoholism: Clin. Exp. Res. 4:104, 1980.

28. Fawcett, J., Clark, D. C., Aagesen, C. A., Pisani, V. D., Tilkin, J. F., Sellers, D., McGuire, M., and Gibbons, R. D.   A double-blind placebo-controlled trial of lithium carbonate therapy for alcoholism. Arch. Gen. Psychiat. 44:248–256, 1987.

29. Freinkel, N., and Arky, R. A.   Effects of alcohol on carbohydrate metabolism in man. Psychosom. Med. 28:551–563, 1966.

30. Fremming, K. H.   Sygdomsrisikoen for Sindslidelser og andre sjaeledige Abnormtilstande i den danske Gennemsnitsbefolkning. Copenhagen: Ejinar Munksgaard, 1947.

31. Fuller, R. K., and Williford, W. O.   Life-table analysis of abstinence in a study evaluating the efficacy of disulfiram. Alcoholism. Clin. Exp. Res. 4:298, 1980.

32. Gerard, D. L., and Saenger, G.   *Out-Patient Treatment of Alcoholism*. Toronto: Univ. of Toronto Press, 1966.

33. Gerard, D. L., Saenger, G., and Wile, R.   The abstinent alcoholic. Arch. Gen. Psychiat. 6:83–95, 1962.

34. Gill, J. S., Zezulka, A. V., Shipley, M. J., Gill, S. K., and Bevers, D. J.   Stroke and alcohol consumption. N. Engl. J. Med. 315:1041–1046, 1986.

35. Glatt, M. M.   An alcoholic unit in a mental hospital. Lancet 2:397–398, 1959.

36. Goodwin, D. W., Crane, J. B., and Guze, S. B.   Alcoholic "blackouts":

a review and clinical study of 100 alcoholics. Amer. J. Psychiat. 126: 191–198, 1969.

37. Goodwin, D. W. Blackouts and alcohol-induced memory dysfunction. In *Recent Advances in Studies of Alcoholism*. Washington, D.C.: U.S. DHEW, 1971.

38. Goodwin, D. W., Crane, J. B., and Guze, S. B. Felons who drink. Q. J. Stud. Alcohol 32:136–147, 1971.

39. Goodwin, D. W. Alcohol in suicides and homicides. Q. J. Stud. Alcohol 34:144–156, 1973.

40. Goodwin, D. W., Schulsinger, F., Hermansen, L., Guze, S. B., and Winokur, G. Alcohol problems in adoptees raised apart from alcoholic biological parents. Arch. Gen. Psychiat. 28:238–242, 1973.

41. Goodwin, D. W., Schulsinger, F., Möller, N., Hermansen, L., Winokur, G., and Guze, S.B. Drinking problems in adopted and nonadopted sons of alcoholics. Arch. Gen. Psychiat. 31:164–169, 1974.

42. Goodwin, D. W., Schulsinger, F., Hermansen, L., Guze, S. B., and Winokur, G. Alcoholism and the hyperactive child syndrome. J. Nerv. Ment. Dis. 160:349–353, 1975.

43. Goodwin, D. W., and Hill, S. Y. Chronic effects of alcohol and other psychoactive drugs on intellect, learning and memory. In *Chronic Effects of Alcohol and Other Psychoactive Drugs on Cerebral Function*. Toronto: Addict. Res. Foundation Press, 1975.

44. Goodwin, D. W., Schulsinger, F., Möller, N., Mednick, S., and Guze, S. B. Psychopathology in adopted and nonadopted daughters of alcoholics. Arch. Gen. Psychiat. 34:1005–1009, 1977.

45. Goodwin, D. W., Schulsinger, F., Knop, J., Mednick, S., and Guze, S. B. Alcoholism and depression in adopted-out daughters of alcoholics. Arch. Gen. Psychiat. 34:751–755, 1977.

46. Goodwin, D. W. Alcoholism and heredity: a review and hypothesis. Arch. Gen. Psychiat. 36:57–61, 1979.

47. Goodwin, D. W. *Alcoholism: The Facts*. New York: Oxford Univ. Press, 1981.

48. Goodwin, D. W. In *DSM-III Case Book*, Spitzer, R. L., Skodol, A. E., Gibbon, M., and Williams, J. B. W. Washington, D.C.: American Psychiatric Association, 1981.

49. Goodwin, D. W. Drug therapy of alcoholism. In *Psychopharmacology*, I: Grahame-Smith, D. G., Hippius, H., and Winokur, G. (eds.) Amsterdam: Excerpta Medica, 1982.

50. Goodwin, D. W. Alcoholism and affective disorders. In *Alcoholism and Clinical Psychiatry*, N.Y. Academy of Medicine. New York: Plenum, 1982.

51. Goodwin, D. W. On defining alcoholism and taking stands. J. Clin. Psychiat. 43:394–395, 1982.

52. Goodwin, D. W. Familial alcoholism. Subst. Alcohol Actions Misuse 4:129–136, 1983.

53. Gorelick, P. B., Rodin, M. B., Langenberg, P., Hier, D. B., Costigan, J.,

Gomez, I., and Spontak, S.   Is acute alcohol ingestion a risk factor for ischemic stroke? Results of a controlled study in middle-aged and elderly stroke patients at three urban medical centers. Stroke 18:359–364, 1987.

54. Guze, S. B. *Criminality and Psychiatric Disorders*. New York: Oxford Univ. Press, 1976.

55. Harper, C., Kril, J., and Daly, J.   Are we drinking our neurones away? Brit. Med. J. 294:534–536, 1987.

56. Harvey, E. B. Schairer, C., Brinton, L. A., Hoover, R. N., and Fraumeni, J. F.   Alcohol consumption and breast cancer. J. Natl. Cancer Inst. 78: 657–661, 1987.

57. Hesselbrock, V. M., Stabenau, J. R., Hesselbrock, M. N., Meyer, R. E., and Babor, T. F.   The nature of alcoholism in patients with different family histories for alcoholism. Neuro-Psychopharmacol. Biol. Psychiatry 6:607–614, 1982.

58. Hill, S. Y., Reyes, R. B., Mikhael, M., and Ayre, F.   A comparison of alcoholics and heroin abusers: computerized transaxial tomography and neuropsychological functioning. In *Currents in Alcoholism*, Sexias, F. (ed.) New York: Grune & Stratton, 1977.

59. Hrubec, Z., and Omenn, G. S.   Evidence of genetic predisposition to alcoholic cirrhosis and psychosis: twin concordances for alcoholism and its biological end points by zygosity among male veterans. Alcoholism: Clin. Exp. Res. 5:207–215, 1981.

60. Hsu, J. J.   Electroconditioning therapy of alcoholics. Q. J. Stud. Alcohol 26:449–459, 1965.

61. Jacobson, R.   The contributions of sex and drinking history to the CT brain scan changes in alcoholics. Psychol. Med. 16:547–559, 1986.

62. Jellinek, E. M. *The Disease Concept of Alcoholism*. New Haven: College & University Press, 1960.

63. Jellinek, E. M.   Phases of alcohol addiction. Q. J. Stud. Alcohol 13: 673–684, 1952.

64. Jones, K. L., Smith, D. W., Ulleland, C. M., et al.   Pattern of malformation in offspring of chronic alcoholic mothers. Lancet 1:1267–1271, 1973.

65. Jones, M. C.   Personality correlates and antecedents of drinking patterns in adult males. J. Consult. Clin. Psychol. 32:2–12, 1968.

66. Kaij, L. *Studies on the Etiology and Sequels of Abuse of Alcohol*. Lund, Sweden: Univ. of Lund, 1960.

67. Keller, M.   On defining alcoholism: with comment on some other relevant words. In *Alcohol, Science and Society Revisited*. Ann Arbor: Univ. of Michigan Press, 1982.

68. Keller, M., McCormick, M., and Efron, V. *A Dictionary of Words About Alcohol*, 2d ed. New Brunswick, N.J.: Rutgers Univ. Center of Alcohol Studies, 1982.

69. Klatsky, A. L., Armstrong, M. A., and Friedman, G. D.   Relations of alcoholic beverage use to subsequent coronary artery disease hospitalization. Am. J. Cardiol. 58:710–714, 1986.

70. Kline, N. S., Wren, J. C., Cooper, T. B., Varga, E., and Canal, O. Evaluation of lithium therapy in chronic and periodic alcoholism. Am. J. Med. Sci. 268:15–22, 1974.

71. Kumar, R., Mac, D. S., Gabrielli, W. F., and Goodwin, D. W. Anxiolytics and memory: a comparison of lorazepam and alprazolam. J. Clin. Psychiat. 48:158–160, 1987.

72. Leevy, C. M., Frank, O., Leevy, C. B., and Baker, H. Nutritional factors in liver disease of the alcoholic. Acta Med. Scand. Suppl. 703:67–79, 1985.

73. Leland, J. Firewater Myths. New Brunswick, N.J.: Rutgers Univ. Center of Alcohol Studies, 1976.

74. Lemoine, P., Harousseau, H., Borteyru, J. P., et al. Les Enfants de parents alcooliques: anomalies observees. Quest. Med. 25:476–482, 1968.

75. Lieber, C. S. Metabolic effects produced by alcohol in the liver and other tissues. Adv. Intern. Med. 14:151–199, 1968.

76. Lieber, C. S., and DeCarli, L. M. An experimental model of alcohol feeding and liver injury in the baboon. J. Med. Primatol. 3:153–163, 1974.

77. DeLint, J., and Schmidt, W. The epidemiology of alcoholism. In Biological Basis of Alcoholism. Toronto: Wiley-Interscience, 1971.

78. Lolli, G., Serianni, E., Golder, G. M., and Luzzatto-Fegiz, P. Alcohol in Italian Culture. Monog. no. 3, Yale Center of Alcohol Studies. Glencoe, Ill.: Free Press, 1958.

79. Ludwig, A. M. On and off the wagon. Q. J. Stud. Alcohol 33:91–96, 1972.

80. Luxenberger, H. Demographische und psychiatrische Untersuchungen in der engeren biologischen Familie von Paralytikerehegatten, Z. Ges. Neurol. Psychiat. 112:331, 1928.

81. MacAndrew, C., and Edgerton, R. B. Drunken Comportment: A Social Explanation. Chicago: Aldine, 1969.

82. MacMahon, S. Alcohol consumption and hypertension. Hypertension 9:111–121, 1987.

83. Maddox, G. L., and Williams, J. R. Drinking behavior of negro collegians. Q. J. Stud. Alcohol 29:117–129, 1968.

84. Marinacci, A. A. Electroencephalography in alcoholism. In Alcoholism, 484–536. Springfield, Ill.: C. C. Thomas, 1956.

85. Mayfield, D. G. Psychopharmacology of alcohol. I. Affective change with intoxication, drinking behavior and affective state. J. Nerv. Ment. Dis. 146:314–321, 1968.

86. McCord, W., and McCord, J. A longitudinal study of the personality of alcoholics. In Society, Culture, and Drinking Patterns. New York: Wiley, 1962.

87. Mendelson, J. H. Biologic concomitants of alcoholism. N. Engl. J. Med. 283:24–32, 1970.

88. Merry, J., Reynolds, C. M., Bailey, J., and Coppen, A. Prophylactic treatment of alcoholism by lithium carbonate. Lancet 2:481–482, 1976.

89. Miller, W. R., and Joyce, M. A. Prediction of abstinence, controlled drinking, and heavy drinking outcomes following behavioral self-control training. J. Consult. Clin. Psychol. 47:773–775, 1979.

90. Morris, H. H., and Estes, M. L. Traveler's amnesia. JAMA 258: 945–946, 1987.

91. Murray, R. M., Clifford, C., and Gurlin, H. M. Twin and alcoholism studies. In Galanter, M. (ed.), vol. 1, chap. 5. Recent Developments in Alcoholism, New York: Gardner Press, 1983.

92. Ogata, M., Mendelson, J., and Mello, N. Electrolytes and osmolality in alcoholics during experimental intoxication. Psychosom. Med. 30:463–488, 1968.

93. Partanen, J., Bruun, K., and Markkanen, T. Inheritance of drinking behavior: a study of intelligence, personality, and use of alcohol of adult twins. In Emerging Concepts of Alcohol Dependence, Pattison, E. M., Sobell, M. B., Sobell, L. C. (eds.), chap. 10. New York: Springer, 1977.

94. Peck, C. C., Pond, S. M., Becker, C. E., and Lee, K. An evaluation of the effects of lithium in the treatment of chronic alcoholism. II. Assessment of the two-period crossover design. Alcoholism: Clin. Exp. Res. 5: 252, 1981.

95. Pemberton, D. A. A comparison of the outcome of treatment in female and male alcoholics. Brit. J. Psychiat. 113:367–373, 1967.

96. Penick, E., Read, M., Crowley, P., and Powell, B. Differentiation of alcoholics by family history. J. Stud. Alcohol 39:1944–1948, 1978.

97. Popham, R. E. Indirect methods of alcoholism prevalence estimation: a critical evaluation. In Alcohol and Alcoholism. Toronto: Univ. of Toronto Press, 1970.

98. Randall, C., Taylor, W., and Walker, D. Ethanol-induced malformations in mice. Alcoholism: Clin. Exp. Res. 1:219–223, 1977.

99. Robins, L. N., Bates, W. M., and O'Neil, P. Adult drinking patterns of former problem children. In Society, Culture, and Drinking Patterns. New York: Wiley, 1962.

100. Robins, L. N., Murphy, G. E., and Breckenridge, M. B. Drinking behavior of young negro men. Q. J. Stud. Alcohol 29:657–684, 1968.

101. Roe, A. The adult adjustment of children of alcoholic parents raised in foster homes. Q, J. Stud. Alcohol 5:378–393, 1944.

102. Romano, J., Michael, M., Jr., and Merritt, H. H. Alcoholic cerebellar degeneration. Arch. Neurol. and Psychiat. 44:1230–1236, 1940.

103. Roueché, B. Alcohol. New York: Grove Press, 1960.

104. Rush, B. An Inquiry into the Effects of Ardent Spirits upon the Human Body and Mind, 6th ed. New York: 1811.

105. Ryback, R. S., Eckardt, M. J., and Felsher, B. Biochemical and hematological correlates of alcoholism and liver disease. JAMA 248:2261–2265, 1982.

106. Sadoun, R., Lolli, G., and Silverman, M. Drinking in French Culture, Monographs of the Rutgers Center of Alcohol Studies, no. 5. New Haven: College and University Press, 1965.

107. Schuckit, M., Pitts, F. N., Jr., Reich, T., King, L. J., and Winokur, G. Alcoholism. Arch. Environ. Health 18:301–306, 1969.
108. Schuckit, M. A., Goodwin, D. W., and Winokur, G. A half-sibling study of alcoholism. Am. J. Psychiat. 128:1132–1136, 1972.
109. Sigerist, H. E. The History of Medicine. New York: M.D. Publications, 1960.
110. Sixth Special Report to the United States Congress on Alcohol and Health. Rockville, Md.: National Institute on Alcohol Abuse and Alcoholism, 1987.
111. Sjögren, T. Genetic-statistical and psychiatric investigations of a west Swedish population. Acta Psychiat. et Neurol., Suppl. 52, 1948.
112. Slater, E. The incidence of mental disorder. Ann. Eugenics 6:172, 1935.
113. Tamerin, J. S., Weiner, S., Poppen, R., Steinglass, P., and Mendelson, J. H. Alcohol and memory: amnesia and short-term memory function during experimentally induced intoxication. Am. J. Psychiat. 127:1659–1664, 1971.
114. Tarter, R. Minimal brain dsyfunction as an etiological predisposition in alcoholism. In Evaluation of the Alcoholic: Implications for Research, Theory and Practice, Meyer, R., Glueck, J., Babor, T., Jaffe, J., Stabenau, J. (eds.), chap. 12. Res. Monog. U.S. DHHS, 1981.
115. Taylor, M. E., and St. Pierre, S. Women and alcohol research: a review of current literature. J. Drug Issues 16:621–634, 1986.
116. Turner, T. B., Bennett, V. L., and Hernandez, H. The beneficial side of moderate alcohol use. Johns Hopkins Med. J. 148:53, 1981.
117. United States Government Public Health Reports 101:593–598, 1985.
118. Vaillant, G. E. The Natural History of Alcoholism. Cambridge: Harvard Univ. Press, 1983.
119. Victor, M., and Adams, R. D. The effect of alcohol on the nervous system. In Proceedings of the Association for Research in Nervous and Mental Disease. Baltimore: Williams & Wilkins, 1953.
120. Victor, M., Adams, R. D., and Collins, G. H. The Wernicke-Korsakoff syndrome. Philadelphia: F. A. Davis, 1971.
121. Victor, M., and Hope, J. M. The phenomenon of auditory hallucinations in chronic alcoholism. J. Nerv. Ment. Dis. 126:451, 1958.
122. Vitols, M. M. Culture patterns of drinking in negro and white alcoholics. Dis. Nerv. Syst. 29:344–391, 1968.
123. Voegtlin, W. L. The treatment of alcoholism by establishing a conditioned reflex. Am. J. Med. Sci., 199:802–809, 1940.
124. Wallgren, H., and Barry, H., III. Actions of Alcohol, vol. I. Amsterdam: Elsevier, 1970.
125. Weissman, M. M., and Myers, J. K. Clinical depression in alcoholism. Am. J. Psychiat. 137:3, 1980.
126. Whitefield, J. B. Alcohol-related biochemical changes in heavy drinkers. Aust. N.Z. J. Med. 11:132, 1981.
127. Willett, W. C., Stampfer, M. J., Colditz, G. A., Rosner, B. A., Hen-

nekens, C. H., and Speizer, F. E. Moderate alcohol consumption and the risk of breast cancer. N. Engl. J. Med. 316:1174–1180, 1987.

128. Woodruff, R. A., Jr., Guze, S. B., and Clayton, P. J. Divorce among psychiatric out-patients. Brit. J. Psychiat. 121:289–292, 1972.

129. Yellowlees, P. M. Thiamin deficiency and prevention of the Wernicke-Korsakoff syndrome. Med. J. Aust. 145:216–219, 1986.

8. Drug Dependence

Definition

Drug dependence refers to the repeated use of a psychoactive drug, causing harm to the user or to others. The World Health Organization recommends this term instead of "addiction" or "habituation" (17). Both these words are still widely used in the literature, however, and require definition.

Addiction is used in two senses, physical and psychological. *Physical addiction* refers to a drug-produced condition characterized by tolerance and physical dependence. *Tolerance* refers to an adaptive biological process in which "increasingly larger doses of a drug must be administered to obtain the effect observed with the original dose." *Physical dependence* refers to "the physiological state produced by the repeated administration of the drug which necessitates the continued administration of the drug to prevent the appearance of the stereotyped syndrome, the withdrawal or abstinence syndrome, characteristic for the particular drug" (21).

Psychological addiction refers to a "behavioral pattern of compulsive drug use, characterized by overwhelming involvement with the use of a drug, the securing of its supply, and a high tendency to relapse after withdrawal" (29). Physical and psychological addiction may overlap but do not necessarily occur together. For example, a person physically addicted to morphine administered for pain in a hospital may have none of the characteristics of psychological addiction. On the other hand, individuals may be psychologically addicted to drugs with little or no tolerance or physical dependence. Except for the opiates and to a lesser extent the barbiturates, no drugs produce striking degrees of both tolerance and physical dependence and therefore meet the two pharmacological criteria for addiction (21).

Habituation refers to dependence on a drug for an "optimal state of well-being, ranging from a mild desire to a craving or compulsion (a compelling urge) to use the drug" (21).

DSM-III-R lists nine symptoms of *psychoactive substance dependence* (Table 8.1). It also includes a category called *psycho-*

Table 8.1 Diagnostic criteria for Psychoactive Substance Dependence (DSM-III-R)

A. At least three of the following:
   (1) substance often taken in larger amounts or over a longer period than the person intended
   (2) persistent desire or one or more unsuccessful efforts to cut down or control substance use
   (3) a great deal of time spent in activities necessary to get the substance (e.g., theft), taking the substance (e.g., chain smoking), or recovering from its effects
   (4) frequent intoxication or withdrawal symptoms when expected to fulfill major role obligations at work, school, or home (e.g., does not go to work because hung over, goes to school or work "high," intoxicated while taking care of his or her children), or when substance use is physically hazardous (e.g., drives when intoxicated)
   (5) important social, occupational, or recreational activities given up or reduced because of substance use
   (6) continued substance use despite knowledge of having a persistent or recurrent social, psychological, or physical problem that is caused or exacerbated by the use of the substance (e.g., keeps using heroin despite family arguments about it, cocaine-induced depression, or having an ulcer made worse by drinking)
   (7) marked tolerance: need for markedly increased amounts of the substance (i.e., at least a 50% increase) in order to achieve intoxication or desired effect, or markedly diminished effect with continued use of the same amount
   Note: The following items may not apply to cannabis, hallucinogens, or phencyclidine (PCP):
   (8) characteristic withdrawal symptoms
   (9) substance often taken to relieve or avoid withdrawal symptoms
B. Some symptoms of the disturbance have persisted for at least one month, or have occurred repeatedly over a longer period of time.

active substance abuse. Individuals fitting this category do not fulfill the criteria for dependence but repeatedly use drugs despite (a) social, occupational, psychological, or physical harm from the substance; or (b) a risk of drug-related accidents. Examples are persons with hepatitis who drink despite warnings from their physicians, or persons who drive while intoxicated. Alcoholism, included by DSM-III-R as a psychoactive substance use disorder, is discussed in chapter 7.

Historical Background

The basic needs of the human race . . . are food, clothing, and shelter. To that fundamental trinity most modern authorities would add, as equally compelling, security and love. There are, however, many other needs whose satisfaction, though somewhat less essential, can seldom be comfortably denied. One of these, and perhaps the most insistent, is an occasional release from the intolerable clutch of reality. All men throughout recorded history have known this tyranny of memory and mind, and all have sought . . . some reliable means of briefly loosening its grip.

                                                        BERTON ROUECHÉ (76)

Whether or not this fully explains the motivation behind the use of intoxicating substances, it is true they have been used by humankind for thousands of years in nearly every part of the world. Except in a few primitive cultures, humans have discovered plants whose juices and powders that on being properly prepared and consumed have caused desirable alterations of consciousness. In ancient times, these substances were widely used in religious ceremonies (as wine is still used in the Catholic mass and peyote by the North American Church), but they also have been used for recreational purposes. As the sciences of medicine and chemistry progressed, the variety of drugs increased and for many years drugs have been diverted from medical use into "illicit" channels as they are today.

The analgesic drugs derived from the poppy plant were particularly susceptible to abuse because they produced euphoria as well as analgesia. Drug "epidemics" have occurred periodically in recent centuries. The introduction of opium into England, commercially promoted as part of the Chinese opium trade, led to widespread abuse of the drug in the nineteenth century. With the introduction of the hypodermic needle during the American Civil War period, the use of morphine for nonmedicinal purposes became widespread, and by the turn of the century large numbers of people were apparently dependent on the drug, taken intravenously or more often as an ingredient of patent medicines. Heroin, the diethylated form of morphine, was introduced around the turn of the century as a "heroic" solution for the opiate problem (a form of substitution therapy, just as methadone is used today).

Another development of the mid-nineteenth century was the introduction of bromides as sedatives. There was an enormous demand for these compounds and a steady increase in their use. With use came misuse, which often resulted in intoxication and psychotic reactions. The bromide problem began to abate in the 1930s as barbiturates and other sedatives became available. The first barbiturate, Veronal, was introduced in 1903 and others appeared in quick succession. The short-acting barbiturates such as pentobarbital and amobarbital became popular in the 1930s and 1940s, their dependence-producing qualities not having been immediately recognized (14).

During the late nineteenth century Westerners discovered botanicals used for mind-altering purposes elsewhere, such as cocaine and hashish (the most potent form of cannabis). Cocaine was found to be useful medically (by Freud, among others), and a number of well-known physicians and surgeons became cocaine addicts.

Ether, also discovered in the nineteenth century, was used recreationally, again often by medical people, as were nitrous oxide and other volatile solvents. Doctors and nurses, presumably because of their access to these drugs and their familiarity with them, were particularly prone to take them, although reliable data regarding drug abuse by physicians at any time in history, including today, are not available. Meperidine, introduced in the 1940s as a "nonaddictive" synthetic narcotic, led to a mild epidemic of abuse among physicians before its addictive potential was appreciated.

In postwar Japan, thousands of young people turned to amphetamines. Drastic measures were required to control the problem, including the establishment of special psychiatric facilities and stringent legal controls.

The most recent epidemic of drug abuse began in the early 1960s when hallucinogens became part of the American middle-class youth culture. It originated with Hofmann's discovery in 1953 that lysergic acid diethylamide (LSD) produced perceptual distortions and other aberrant mental phenomena. Again, physicians were involved in the introduction of the drug (41). It was first used on the U.S. East and West coasts experimentally because it produced symptoms resembling those seen in psychosis. It was also used therapeutically, usually by psychoanalysts who felt the drug would dissolve "repressions." Because LSD is easy to synthesize, it became widely available and has been used, at least experimentally, by several million people, the majority of them of young age and from the middle or upper-middle class.

As mind-altering drugs became widely publicized in the 1960s, more and more young people experimented with them. In the early adolescent group, glue-sniffing and inhalation of other volatile solvents was apparently widely practiced, whereas persons in late adolescence and then early twenties, particularly on college campuses, experimented with LSD and other synthetic hallucinogens such as dimethyltryptamine (DMT) and other derivatives of the indole and catecholamines that produced psychedelic effects. Starting in the mid-1970s, the use of LSD and other psychedelics began to decline and has steadily continued to decline every year since then (23). Meanwhile, increase in the use of amphetamines produced a generation of speed freaks who injected methamphetamine intravenously. Downers such as barbiturates and other sedative-hypnotics also became popular among young people in the late 1960s. Cocaine became available in suburbia and on college campuses, although, as in the case of other street drugs, often the "cocaine" people thought they were obtaining was either adulterated or completely absent. This was also the case with mescaline, the active principal of peyote (53).

Apparently because marihuana also enhances sensory processes (54) and also is abundantly available, it became the most widely used illicit LSD-like drug in America during the 1960s and 1970s. Marihuana, the "tobacco" of the hemp plant, has grown wild in America since colonial times or before, but until the past decade was used recreationally only by small minorities such as jazz musicians in the 1930s and Mexican immigrant workers (6). Widely condemned in the 1930s as a dangerous drug, marihuana has been increasingly accepted in Western countries as a benign if not completely innocuous substance, and there has been increasing pressure to "decriminalize" the drug if not actually make it legally available. Marihuana and its potent relative, hashish, are still, however, illegal in most countries, many of which are signatories to a U.N. treaty prohibiting its sale.

Although marihuana and the hallucinogens were being widely used in middle- and upper-middle-class America in the 1960s, heroin—the fast-acting, potent form of morphine—became a serious medical and legal problem, involving mainly lower-class black urban men. In the early 1980s it was estimated that New York City alone had more than one hundred thousand opiate addicts. Because of the expense of heroin, its use led to widespread criminal activities, including corruption of police authorities. Deaths resulted from overdosage, allergic responses, or medical complications following unsterile intravenous injection.

The heroin "epidemic" was limited somewhat by the high price of the drug and its relative unavailability in noncoastal cities. However, during the later stages of the American involvement in the Vietnam War, pure grades of opium became widely available to American servicemen in Vietnam. One study suggests that nearly half the American soldiers in Vietnam during 1971 experimented with opium or its derivatives and that 20 percent were frequent users (19).

Epidemiology

Epidemiological surveys of drug use involve not only the usual un-
reliability expected in prevalence studies but also problems peculiar to drug studies, especially the rapidly changing patterns of use.
It is generally agreed, for example, that use of marihuana and cocaine has increased vastly since 1970, but data from earlier periods
are lacking for comparative purposes.

One fact is certain: Leaving aside caffeine-containing beverages,
alcohol and tobacco dominated the recreational drug scene in the
United States until the 1960s. Opposition to the Vietnam War
and the sexual revolution in the 1960s were accompanied by an
explosion of illicit drug use, ranging from a rediscovery of the
products of the poppy and cannabis plants to new "designer"
drugs that could be synthesized in basements by people with a
modicum of technical knowledge—drugs not even illegal until
laws could be passed to make them so (15). The use of these
drugs, singly and in combination, has produced a huge public
health problem in Western countries. Among young people in
particular, polydrug use has led to a spectrum of bizarre reactions
as confusing to physicians as to worried parents.

It now appears that drug abuse in general is leveling off and
that there is a modest decline in alcohol use and a more dramatic
reduction in smoking. A great many people, however, have been
exposed to a wide variety of illicit drugs. According to government
figures (61, 62, 79), by 1988 some 60 million Americans had tried
marihuana and some 25 million used it regularly. At least 20 million Americans have tried cocaine and 5 million are regular users.
There are about a half million heroin users—perhaps half of them
regular users. More than $1 billion is spent on illicit drugs. An-
other $60 billion are lost in drug-related crime, health problems,
and lowered productivity. Intravenous drug users are second only
to homosexuals as a risk group for acquired immune deficiency
syndrome (AIDS).

The image of drug abusers has changed. Socialites, athletes, and

Hollywood celebrities have been arrested in increasing numbers for possession of cocaine. Marihuana has been partially decriminalized in some states: selling it is still a crime, but possession of small amounts is not. Nevertheless, law enforcement devotes heavy resources to combating marihuana use. In 1985 local police made 451,138 marihuana arrests, most for simple possession. These comprised half of *all* drug arrests, despite recognition that heroin and cocaine are greater menaces (65). The popularity of crack—a hazardous derivative of cocaine—in affluent suburbia as well as the ghetto seems one of the most serious perils in this very rapidly changing drug scene, whereas LSD and the other hallucinogens have declined greatly in use since the 1970s.

Drug Categories

The Opiates

Morphine is typical of this group. Other opiates include heroin, dilaudid, codeine, and paregoric. Meperidine (Demerol) is a synthetic analgesic structurally dissimilar to morphine but pharmacologically similar. Heroin is the only one of these substances not available for medical use in the United States.

In usual therapeutic doses, morphine produces analgesia, drowsiness, and a change in mood often described as a sense of well-being. Morphine is also a respiratory depressant (often the cause of death in cases of overdose); it produces constriction of the pupils, constipation, and occasionally nausea and vomiting.

Most heroin or morphine addicts take opiates intravenously, which produces flushing and a sensation in the lower abdomen described as similar to orgasm, known as a "rush." Initial increased activity is followed by a period of dowsiness and inactivity, which is called "being on the nod." During this time opium dreams occur.

Tolerance develops to most of morphine's effects, including euphoria. Libido declines; menstruation may cease. Some addicts who are able to obtain an adequate supply of opiates dress properly, maintain nutrition, and discharge social and occupational obligations without gross impairment. Other addicts become socially disabled, often because the expense of drugs may lead to theft, forgery, prostitution, and the sale of drugs to other users.

The first symptoms of opiate withdrawal usually appear eight to twelve hours after the last dose (Table 8.2). These include lacrimation, rhinorrhea, sweating, and yawning. Thereafter, the addict may fall into a fitful sleep known as the yen from which he or she awakens after several hours, restless and miserable. Additional signs and symptoms appear, reaching a peak between forty-eight and seventy-two hours after drug discontinuation. These include dilated pupils, anorexia, restlessness, insomnia, irritability, gooseflesh, nausea, vomiting, diarrhea, chills, and abdominal cramps. These manifestations subside gradually. After seven to ten days, most signs of abstinence have disappeared, although the patient may continue to experience insomnia, weakness, and nervousness.

In addition to acute withdrawal symptoms, opiates may produce a protracted abstinent syndrome lasting several months (39). Body temperature and respiratory rate may be below normal. The patient may have trouble sleeping. More important, the craving for opiates may persist for several months after the last dose of heroin. This probably contributes to the high recidivism rate in opiate users. Another reason for recidivism has been attributed to a phenomenon called the conditioned abstinent syndrome (13, 57, 68). Detoxified patients who return to the environment in which they use drugs may experience craving for the drug as well as physical manifestations of abstinence because of conditioned stimuli (seeing old friends, familiar places, and so forth).

To ameliorate withdrawal symptoms, oral methadone can be substituted for any opiate. Methadone can be discontinued gradually over a period of one or two weeks. Clonidine, an alpha-adrenergic agonist given for hypertension, also alleviates the opiate withdrawal syndrome, suggesting that catecholamines are involved in opiate withdrawal. Clonidine has one advantage over methadone: It allows the patient to become narcotic-free (39, 46).

The course of opiate addiction varies depending on the population studied. In Vaillant's twelve-year follow-up of opiate addicts originally treated at the federal addiction treatment center at Lexington, 98 percent had returned to the use of opiates within twelve months after release (94). On the other hand, a study of Vietnam veterans who used opium and its products extensively during their tour in Vietnam revealed that less than 2 percent continued using the drug in the year after their return to the United States (19). This suggests that circumstances of exposure and availability are important factors in maintaining use.

In America, most heroin addicts are men and, at least in inner cities on the two coasts, a majority are nonwhite. There is some evidence that heroin users from this population who do not die as a result of drugs or criminal activities may spontaneously stop using the drug after their mid-thirties. School problems such as delinquency and truancy frequently antedate heroin use (8).

Even in environments where addiction is common, there seems to be a sizable number of occasional or controlled users of heroin, referred to as chippers. Although well known in the street, they are rarely mentioned in the professional literature, where they are generally assumed to be people in a transition stage leading to either compulsive use or abstinence. In fact, the evidence suggests that some people may use heroin occasionally and without signs of addiction for many years (71). In the case of any street drug, however, the purity of the substance must be taken into account. Chippers may be able to control their heroin use because the heroin is of low potency.

Except perhaps for sociopathy, there is little evidence that opiate users have other diagnosable psychiatric illnesses, although many of them take other drugs such as alcohol and barbiturates. Multiple drug use, based on observation in methadone clinics (see later discussion), is considered a bad prognostic sign in heroin addicts.

The mortality from opiate withdrawal is negligible despite the unpleasantness of the withdrawal syndrome. Deaths occurring among addicts are mainly due to infection, overdose, and apparent hypersensitivity reactions, possibly from diluents such as qui-

Table 8.2 Abstinence signs in sequential appearance after last dose of opiate in patients with well-established parenteral habits

Grades of Abstinence	Signs (observed in cool room, patient uncovered or under only a sheet)	Hours After Last Dose					
		Morphine	Heroin	Meperidine	Dihydro-Morphinone	Codeine
Grade 0	Craving for drug Anxiety	6	4	2–3	2–3	8
Grade 1	Yawning Perspiration Lacrimation Rhinorrhea Yen sleep	14	8	4–6	4–5	24
Grade 2	Increase in above signs plus: Mydriasis Gooseflesh (piloerection) Tremors (muscle twitches) Hot and cold flashes Aching bones and muscles Anorexia	16	12	8–12	7	48
Grade 3	Increased intensity of above plus: Insomnia Increased blood pressure	24–36	18–24	16	12	—

Increased temperature (1–2)					
Increased respiratory rate and depth					
Increased pulse rate					
Restlessness					
Nausea			16	—	
Grade 4	Increased intensity of above plus:	36–48	24–36	—	—
	Febrile facies				
	Position—curled up on hard surface				
	Vomiting				
	Diarrhea				
	Weight loss (5 lb. daily)				
	Spontaneous ejaculation or orgasm				
	Hemoconcentration				
	leukocytosis, eosinopenia,				
	increased blood sugar				

Note: Not all signs are necessary to diagnose any particular grade. Racemorphan (Dromoran) and levorphanol (Levodromoran), although three times and six times as strong as morphine sulfate, show same time curve as morphine sulfate. Similarly paregoric, laudanum, and hydrochlorides of opium alkaloids (Pantopon), depending on their relative content of morphine. Adapted from Am. J. Psych. (5).

nine. Unsterile injections may produce septicemia, hepatitis, sub-acute bacterial endocarditis, tetanus, and AIDS. Uncertain dosage is often the cause of death from opiates. The powders sold on the street are highly variable in potency. Accustomed to the successive dilutions of the drug as it passes through the distribution channels, addicts may unexpectedly encounter relatively pure heroin and consequently overdose.

In studies in metropolitan Chicago, Hughes and his colleagues have found that heroin addiction occurs in epidemics, that heroin use spreads from peer to peer, and that newly involved cases are more "contagious" than chronic cases. Microepidemics—five to fifteen new cases over a five-year period—are seen in some locales, and macroepidemics—more than fifty new cases over a five-year period—in others. Hughes and his associates believe macroepidemics are most likely to occur in areas of high unemployment and crime, deterioration in housing, poor or absent neighborhood leadership, a large recent influx of poor people, and the many other expressions of urban blight (26). Their observations parallel those of Leighton et al. (35), who found a positive correlation between social disorganization and opiate use.

Hughes has provided some evidence that an outreach effort by community action groups can be used to abort an epidemic of heroin addiction. However, this was accomplished in a predominantly white neighborhood and was limited to microepidemics (27).

The superiority of one treatment over another with regard to treating opiate addiction has not been established. Some years ago, methadone maintenance was introduced as a long-term treatment for narcotic users (23). Methadone is given in gradually increasing doses over a period of weeks until a high degree of cross-tolerance to all opiates is established. The euphoric effects of intravenous narcotics are blocked. Some previous drug users have become stabilized as productive citizens with this treatment. Apparently methadone alone is not enough for this outcome in many addicts; a complete program of rehabilitation is necessary.

The original methadone program of Dole and Nyswander (22)

was considered impressively successful but has been criticized in that multiple drug users and other presumed poor prognosis cases were not accepted in the program. The approach also has been challenged, in that it involves substituting one addiction for another, and has led to wide dissemination of methadone through illicit channels (63). The consensus at present is that high doses of methadone are usually required for the program to be effective.

Nevertheless, in many cities, methadone has been given credit for ending the criminal activity of a high percentage of heroin addicts. In a four-year follow-up of methadone clients, Dole et al. (23) found that the majority were productively employed, living as responsible citizens, and supporting families. Ninety-four percent no longer engaged in criminal activity to support their habit.

Self-help groups made up of former addicts have stimulated enthusiastic responses from some participants, but their efficacy in general remains unknown.

Opiate research has led to exciting findings in the past decade. Injecting opiate compounds labeled with radioactive tracers into rats, investigators discovered that specific opiate receptors exist on the surfaces of particular cells in the brain stem and that the attachment of opiates to these cells could be prevented by morphine antagonists. This led to the discovery of endogenous opiates (endorphins) in pituitary gland secretions. The history of this remarkable work, involving several different laboratories, can be found in two review articles (84, 89).

Sedative-Hypnotics

Sedatives produce calmness; hypnotics produce sleep. Both share similar pharmacological actions, hence the term "sedative-hypnotics." In higher doses, drugs marketed as sedatives also produce sleep. Conversely, hypnotics prescribed in lower doses produce calmness. The most popular class of sedative-hypnotics are the benzodiazepines (discussed in the next section). Benzodiazepines are also called tranquilizers or "minor" tranquilizers, in contradistinction to "major" tranquilizers such as the antipsychotic drugs.

The term "tranquilizer" originated from the observation that benzodiazepines would "tranquilize" wild animals—make them more tame and manageable—without rendering them unconscious (35). Benzodiazepines are distinguishable from barbiturates and older sedative-hypnotics such as choral hydrate, in that animals and people can easily be aroused from benzodiazepine-produced sleep, whereas this is more difficult with the older drugs.

This section will confine itself to barbiturates and newer synthetic hypnotics that also produce dependence and withdrawal symptoms, for example, glutethimide (Doriden), methyprylon (Noludar), and ethchlorvynol (Placidyl).

In contrast to opiates, barbiturates are usually taken orally. The short- and intermediate-acting barbiturates are most frequently used for nonmedical purposes. With both acute and chronic intoxications as well as the withdrawal syndrome, the clinical picture resembles that produced by alcohol. An intoxicated patient may show lethargy, difficulty in thinking, poor memory, irritability, and self-neglect. Nystagmus, ataxia, and a positive Romberg sign are often present (14).

Withdrawal symptoms may occur when patients take over 400 mg of barbiturates per day (20). The patient becomes anxious, restless, and weak during the first twelve to sixteen hours after withdrawal. Tremors of the hands are prominent; deep tendon reflexes are hyperactive. During the second or third day, grand mal seizures may occur, sometimes followed by delirium. Deaths have been reported. The syndrome clears in about a week (14). Long-acting barbiturates such as phenobarbital rarely if ever produce major withdrawal symptoms.

Withdrawal manifestations may be prevented or treated with large doses of a short-acting barbiturate such as secobarbital (800 to 1000 mg per twenty-four hours in divided doses) with gradual reduction of the dose over a period of seven to ten days (20, 69). If a patient shows evidence of intoxication with the first or second dose (200 mg each), the diagnosis of barbiturate abuse should be questioned in view of the absence of tolerance. Because of cross-tolerance between barbiturates and other sedatives and hypnotics, withdrawal from these drugs may also be accomplished with a short-acting barbiturate.

With regard to individuals particularly vulnerable to barbiturate abuse, there is little agreement about a so-called addictive personality other than the application of ambiguous and poorly defined terms such as "passive-dependent personality." It is often said that drug dependence reflects an underlying mental disorder, but the nature of the disorder is usually not specified. As with alcoholism, barbiturate dependence leads to physical and social complications that may involve loss of health, family, and job; accidents; and traffic offenses. Also, there is evidence that barbiturates compete with alcohol in their propensity to promote crimes of violence (62).

Little is known about the natural history of barbiturate dependence but clinical experience indicates that, in common with alcoholism, the course is characterized by relapse and chronicity. There is no specific treatment, and it is not known whether individual or group psychotherapy or drug substitution (e.g., prescribing tranquilizers) is helpful. Where barbiturates are used to counteract the effects of stimulant drugs, such as amphetamines it is imperative that the multiple drug abuse pattern be attacked. Particularly in young people today, multiple drug abuse seems to have a mutually reciprocating, cyclical pattern of stimulation-sedation. Some individuals in this group try to achieve both effects simultaneously. Barbiturates and other sedatives also are frequently used with alcohol and opiates. These compounds have a synergistic effect with alcohol, and "accidental" suicides have been attributed to their combined use.

It may be unwise to prescribe barbiturates to depressed patients unless dosage is carefully controlled. Patients often accumulate large amounts of drugs by hoarding. Unintentional death may result from slow absorption rates because large quantities of barbiturates in the stomach diminish gastric and intestinal function (14). Users not getting the desired effect in what seems a long time continue to take tablet after tablet until they are unconscious. In the process, they ingest a lethal dose.

Like alcoholics, sedative-dependent individuals often deny the extent of their drug problem. When use of the drug is denied or minimized, it may be difficult to distinguish the condition from depression, organic brain syndrome, or other psychiatric or neurological disorders. When a patient has a grand mal seizure, it should alert the physician to possible sedative dependence. Seizures are most likely to occur one or two days after withdrawal, although in the case of certain newer synthetic hypnotics, such as glutethimide (Doriden), seizures may occur after a longer period of time. Occasional deaths have been reported during withdrawal, but it is often unclear whether the patient had an intercurrent medical condition at the time of death. Ordinarily, the withdrawal syndrome is self-limited.

Two members of the minor tranquilizer and sedative group appear to have a particularly high potential for abuse. One is meprobamate, the first minor tranquilizer to become available; the other is methaqualone. Both have a risk potential roughly equivalent to that of the short-acting barbiturates. Methaqualone abuse became common among adolescents reaching almost epidemic proportions by the early 1980s (28, 68). One reason for the increase apparently was the widespread belief that methaqualone is an aphrodisiac.

In part, abuse of barbiturates and other sedative-hypnotics is iatrogenic. Unlike most of the other drugs discussed in this chapter, these compounds are primarily available by prescription (14). In prescribing such drugs, there is a particular responsibility to administer the minimum necessary to control the patient's symptoms. Generally, prescriptions should be nonrefillable.

Phenobarbital is useful in detoxifying sedative-hypnotic abusers. The drug is long acting and produces a gradual cessation of withdrawal symptoms: 30 mgs of phenobarbital is roughly equivalent to 100 mgs of the short-acting barbiturates. Often there is uncertainty about the type and quantity of the sedative-hypnotics the patient has been using. In this event, repeated doses of phenobarbital may be given in smaller amounts to prevent withdrawal (86, 93).

Benzodiazepines

Benzodiazepines are the most widely prescribed class of medication in the Western world. At least a dozen such drugs are marketed in the United States and perhaps thirty or more throughout the world. Compared to the sedative-hypnotic drugs described earlier, benzodiazepines have the following advantages: (a) they relieve anxiety without necessarily reducing alertness, impairing coordination, or interfering with normal thinking processes; (b) the regularly prescribed doses provide little or no euphoria; (c) they are good muscle relaxants; (d) they are the safest, most effective drugs for treating alcohol withdrawal and other agitated states produced by many street drugs such as LSD (35). Benzodiazepines can be given to intoxicated patients in an emergency room without fear of lethal reactions, although caution should always be exercised. One of the newer benzodiazepines, alprazolam (Xanax), may relieve depression and panic attacks as well as anxiety. Benzodiazepines are the best drugs for insomnia. Whether, like barbiturates, they lose their effectiveness in a short time is not clear, but the shorter-acting benzodiazepine hypnotics such as triazolam (Halcion) and temazepam (Restoril) leave few aftereffects the next morning. After prolonged use, patients often experience increased insomnia for a few days—possibly a rebound from the previous hypnotic effects—but the difficulty seems to be short-lived. Finally, these drugs have one great advantage over the barbiturates and other hypnotics of the past: it is almost impossible to commit suicide by taking an overdose. Very rarely, deaths are reported from overdose, but almost always only when accompanied by other sedative agents or alcohol (35).

Reports of a benzodiazepine withdrawal syndrome have appeared for many years. Although millions of people regularly use these drugs, serious withdrawal symptoms appear uncommon (25). The symptoms range from panic and hallucinations to disorientation and grand mal seizures. Benzodiazepine users often combine the drug with alcohol or other drugs, and it is usually not clear whether the benzodiazepine alone is responsible for the abstinence syndrome (29). There has been considerable interest in abstinence syndromes associated with low doses of benzodiazepines taken over a course of years (70, 73).

Nevertheless, in most animal and human studies designed to predict dependence liability, benzodiazepines rank low (39). In one study, subjects were given a choice between 10 mgs of diazepam or placebo; they chose the placebo (45). The same subjects preferred a 5-mg dose of dextroamphetamine to placebo. Nevertheless, physicians should be cautious in prescribing the drug for long periods, even in low dosage. Inevitably, some patients will combine them with other substances. Also, abrupt withdrawal of the drugs after relatively high dosages may produce severe anxiety and psychotic symptoms. Alprazolam (Xanax), an effective drug for anxiety, has the disadvantage of producing some degree of physical dependence, even when taken in moderate dosage. Withdrawing a patient from alprazolam should be done slowly, reducing the dosage by as little as 10 percent per week (59). Geriatric patients are particularly sensitive to benzodiazepines and should be given smaller doses than younger patients.

Amphetamines

This group of stimulants includes d-amphetamine (Dexedrine), d-l-amphetamine (Benzedrine), and methamphetamine (Desoxyn, Methedrine). Their effects include elevated mood, increased energy and alertness, decreased appetite, and slight improvement in task performance. Amphetamines may be taken orally or injected intravenously.

Drug-induced psychosis is common among individuals taking large doses of amphetamines (over 50 mg per day). Such psychosis may resemble paranoid schizophrenia with persecutory delusions occurring despite a clear sensorium. Hallucinations are common. Usually the psychosis subsides one or two weeks after drug use stops. When an apparent amphetamine psychosis persists, a hidden source of drugs or the diagnosis of schizophrenia should be considered.

No specific amphetamine withdrawal syndrome has been described; however, users complain of fatigue and depression after discontinuing the drug, and these have been temporally associated with a decrease in the excretion of 3-methoxy-4-hydroxphenylglycol (MHPG), a metabolite of norepinephrine (66). Tolerance may be striking. Speed freaks sometimes take as much as 1 or 2 gm per day (the usual therapeutic dose is 5 to 15 mg per day).

Since their introduction in the early 1930s, amphetamines have been prescribed by doctors for a variety of conditions, including depression, obesity, and narcolepsy. Starting in the 1960s increasingly more of the substance found its way into illegal distribution channels and became available on the street. Amphetamine use declined in the 1980s, partly because of doctors' reluctance to prescribe the drug as well as the growing popularity of cocaine.

Concern about the increased use of amphetamines and other stimulants led to a marked curtailment of their production and availability. Cocaine, considered by many in the 1970s as a relatively benign drug, substantially replaced amphetamines as a street drug. In the opinion of Hollister and others (39), the trade-off was a loss. First, the experience of the 1980s indicates that cocaine is more dependence producing and dangerous than the amphetamines. Second, the previous beneficial use of amphetamines for medical purposes were virtually denied. In some countries, amphetamines were removed from the market. Yet the evidence was strong that these and similar drugs are useful for treating patients with narcolepsy, attention-deficit disorders, and depression in selected patients. "All these uses," wrote Hollister, "became obsolete simply because a few individuals got into trouble after they used doses one or two orders of magnitude greater than therapeutic doses" (39). Supporting his point in a general way is the 1987 report by the National Institute of Drug Abuse that nonmedical use of prescription drugs as an individual's only form of drug abuse is relatively rare; only 3 percent of the total population had used prescription drugs nonmedically and not used marihuana or any other illicit drug (62).

Cocaine

The United States has experienced two major epidemics of cocaine abuse. The first occurred in the last decades of the nineteenth century when physicians, in particular, became addicted to
the drug (thinking, as did Freud at first, that it was relatively benign and nonaddictive). The second epidemic began in the mid-
1970s and showed no signs of abating ten years later. According to
a 1987 report by the National Institute of Drug Abuse (65),
cocaine remained a major drug of abuse throughout the country.
In New York deaths because of cocaine increased 46 percent between 1985 and 1986. In Philadelphia, hospital admissions for
problems stemming from cocaine use exceeded those related to
heroin. Among defendants charged with major offenses, urinalysis
indicated that the number of persons testing positive for cocaine
had more than tripled from 1984 to 1987. In some cities, cocaine
had become the leading cause of drug-related emergency room
visits, surpassing even those involving alcohol.

Increasing numbers of white, employed, middle-class people in
their late twenties and thirties who had never before sought psychological help now sought treatment for cocaine dependence. By
1985 a cocaine hot line in Washington was receiving more than
14,000 calls daily (32, 95). Many of the callers were college grad-
uates, professionals, and business executives. Half had incomes
over $25,000 per year and a cocaine habit costing $600 a week. Dependence on the drug was intense. Nearly all callers said they
could not turn the drug down when available. Most said cocaine
had become more important to them than food or sex. They had
physical complaints: headaches, sinus infections, sexual malfunc-
tioning. Some had seizures or became paranoid.

The demographics of the epidemic changed rapidly as the country moved into the second half of the 1980s (80). Use among
women, minority groups, lower-income groups, and adolescents increased. Cocaine problems infiltrated the small towns of the Middle West. Once called the champagne of drugs, cocaine became
less expensive, freebasing (smoking cocaine) became popular, and

cheap freebase, known as crack, invaded the slums as well as suburbia. By the mid-1980s, cocaine was universally considered a dangerous drug (69).

Cocaine resembles amphetamines in many ways, but whereas
amphetamines are synthetic drugs, cocaine is of botanical origin,
deriving from shrubs grown in the Andes. Its effects last for a
shorter period than do those of amphetamines. Like amphetamines,
cocaine is a powerful stimulant of the nervous system, producing
euphoria, increasing alertness, and suppressing fatigue and bore-
dom. After regular, heavy use of the drug, a withdrawal syndrome
may occur, in which formication (sensation of bugs crawling under the skin) is frequent together with depression and lassitude
(95).

Heavy use of cocaine may produce a schizophreniform psychosis
resembling paranoid schizophrenia. Occasional users sometimes report a profound dysphoria, persisting for several days after they
have used the drug.

Until a few years ago, the cocaine most often smoked was cocaine hypochloride, the form most widely available on the street.
But in the mid-1970s Americans adopted a method similar to the
South American practice of smoking cocoa paste (base)—a potent
extract from cocoa leaves. Users "free" street cocaine from its salts
and cutting agents by a chemical process—potentially hazardous
because it involves heating flammable solvents. The purified cocaine base that is left, mainly cocaine alkaloid, is highly potent
and is believed by authorities to be very hazardous. It increases
heart rate and blood pressure (97); produces tremors, anorexia,
insomnia; sometimes causes hallucinations (82), paranoia, and depression; and even results in occasional cardiac arrest (56) and
lung damage (20). Nevertheless, according to a 1985 National
Household Survey, 20 percent of cocaine users were using freebase despite its known dangers (62).

Phenothiazines may be used to treat the symptoms of amphetamine and cocaine psychoses. Antidepressants may be used to
treat associated depressive symptoms (30). Intravenous diazepam
can be used to control seizures and propranolol to block other effects of amphetamine and cocaine overdosage (39). Treatment

for control of addiction tends to be nonspecific and to rely heavily on self-help groups and behavior modification (64). Lithium carbonate has been reported to block the effects of stimulants, but its value in the treatment of stimulant addiction awaits further study (39).

Hallucinogens

Perceptual distortions are the primary effect of hallucinogenic drugs, which include LSD, mescaline, psilocybin, and a family of compounds resembling either LSD or the amphetamines chemically (e.g., DMT and STP [2,5-dimethoxy-4-methylamphetamine]). Except for mescaline and psilocybin, most hallucinogens are synthetic. Because LSD has been the most widely used and studied hallucinogen, our discussion will focus on it.

Together with perceptual distortions, including illusions and hallucinations, LSD trips produce altered time sense, disturbed judgment, and sometimes confusion and disorientation (1, 39). Some LSD trips are unpleasant. LSD users may be brought to emergency rooms for bad trips characterized by panic and sometimes persecutory delusions. There are reports of prolonged or flashback effects (a resurgence of the drug effect days, weeks, or months subsequent to the end of a trip). Although flashbacks have not been studied systematically, they are reported so commonly that some authorities believe individuals do experience them. A number of deaths have been attributed to LSD, primarily from suicide or homicide, although the causal connection remains unclear (32, 58).

Repeated use of LSD results in marked tolerance. No withdrawal syndrome has been reported. The drug usually produces euphoria, but dysphoria may occur. No deaths have been reported from LSD overdosage. Retinal damage may occur from gazing directly at the sun.

The use of LSD peaked in the 1970s and has been in steady decline since then. In 1985 only 11 percent of young adults had ever used hallucinogens, a drop from 20 percent in 1982 (62).

Treatment for acute adverse reactions to LSD consists of reassurance and, if this fails, tranquilizers. Diazepam appears to be as effective as phenothiazines and is somewhat safer because it is less likely to produce hypotension.

Hallucinogens have complicated the problem of diagnosing psychiatric disorders among young drug users. Prolonged psychotic episodes are probably more often attributable to schizophrenia or other psychiatric illness than to LSD toxicity. To determine whether a patient has a psychiatric illness that is not associated with drug abuse, the patient must be drug free for a lengthy period.

Phencyclidine

Angel dust (PCP) has become an important drug of abuse in recent years (30, 55). Apparently it has caused a number of deaths (30, 49, 64). Though PCP is obtainable as a veterinary anesthetic (Sernylan), most of the illegal supply is manufactured in home laboratories. The drug is available in many forms—tablet, powder, leaf, rock crystal—and can be ingested, snorted, smoked, or injected. It is often sold under false guise as THC (tetrahydrocannabinol), LSD, or mescaline. Marihuana and other street drugs are often adulterated with PCP. It has been called the "universal adulterant" (16).

Moderate doses of PCP may produce bizarre behavior that is often accompanied by a blank stare. Users become confused, combative, stuporous, or comatose. Large doses can produce seizures and very large doses respiratory depression. Myoclonic jerks are common. Deaths are believed to be caused by hypertensive crises (37). One way to distinguish PCP intoxication from that of other street drugs such as LSD and mescaline is to inspect the pupils. Patients intoxicated with most hallucinogens have dilated pupils; with PCP the pupils are normal or small (2). Ataxia, nystagmus, muscular rigidity, and normal or small pupils in a combative or stuporous patient suggest the diagnosis of PCP intoxication (16).

Phencyclidine is a "dissociative" anesthetic. It induces a cataleptic state in which pain is not perceived; the individual's eyes are wide open, but he or she is unconscious and amnesic for the experience. A noteworthy feature of PCP is the behavioral discontrol that can produce a variety of bizarre activities or violence that, in turn, results in encounters with police, lawsuits, unusual accidents, strange suicides, and impulsive homicides. The user is often amnesic about these events, and may remember only the pleasant floaty euphoria the drug produces at low doses. Some of the reports are reminiscent of LSD descriptions: slowing of time, feelings of power, loss of appetite, illusions and pseudohallucinations. Flashbacks apparently occur even more frequently with PCP than LSD. Overdose with PCP is manifested by a prolonged eyes-open coma, often accompanied by a series of convulsions (67).

If the patient recovers from the overdose, a period of agitated delirium may ensue. This is marked by paranoid delusions, hallucinatory experiences, disorientation, and restlessness. If the dose of PCP was insufficient to induce coma, then the toxic delirium may be the presenting clinical picture. It may last up to a week after the last dose (3).

These patients often are admitted to psychiatric hospitals with a diagnosis of acute paranoid or catatonic schizophrenia. They are seen so frequently at present that acute schizophrenics should have a routine urine test for PCP and amphetamines (28).

Phenothiazines, the usual drug class of choice for agitated and uncontrollable patients, can cause severe postural hypotension and may increase muscle rigidity in patients intoxicated with PCP. Intramuscular benzodiazepines may be a better choice. Talking down a patient intoxicated with PCP may not be as effective as with LSD and other hallucinogens and can instead overstimulate the patient (16).

Marihuana

Cannabis sativa (hemp) contains chemicals called cannabinoids, one of which, delta-9-tetrahydrocannabinol (THC), is believed to be the main psychoactive constituent of marihuana.

The effects of THC depend on dosage, and they range from mild euphoria with small doses to LSD-like effects with high doses. Most marihuana grown wild in the United States contains insufficient amounts of THC to produce measurable psychological effects. Cannabis imported from Mexico, South America, and India often contains sufficient THC to produce marked euphoria and, depending on the plant preparation used, paranoid delusions and other psychotomimetic effects. But THC deteriorates at the rate of 5 percent per month at room temperature; even imported cannabis eventually loses its potency (36). However, cannabis in use today tends to have a much higher THC content than that available in the mid-1970s (55).

There are two generally available cannabis preparations: marihuana and hashish. The former consists of a mixture of plant products (leaves, stems, etc.). The latter consists of the resin of the female plant and is more potent.

In parts of the world where concentrated cannabis preparations have been used by millions of people for hundreds of years, constant use is said to be associated with serious medical and psychiatric illnesses (9).

However, studies conducted in Jamaica, Costa Rica, and Greece—countries where cannabis has been used regularly for many years—failed to reveal any serious pathology associated with the use of the drug (77). The subjects were lower-class laborers and no conclusions could be drawn about deleterious effects in other societies and in other classes. The studies did report a decrease in respiratory capacity and minor hematological abnormalities among users. The authors warned against extrapolating their findings to other societies.

Acute psychotic episodes apparently have resulted from cannabis use, but they have been infrequent (31). However, a Swedish study (3) published in late 1987 reported a strong correlation between cannabis use and schizophrenia, defined by DSM-III criteria. The study involved a cohort of nearly fifty thousand Swedish military conscripts. Those who reported using cannabis on more than fifty occasions were six times more likely subsequently to receive a diagnosis of schizophrenia than were nonusers. The relative risk increased with increased consumption. The study did not demonstrate that cannabis use *caused* schizophrenia; cannabis consumption might, on the contrary, be caused by an emerging schizophrenia. Nevertheless, the association between marihuana use and schizophrenia was the strongest yet reported. Psychiatrists have also described a condition among chronic users of marihuana called the amotivational syndrome, characterized by lassitude, apathy, and lack of ambition (54, 78). The evidence for a causal relationship with cannabis remains anecdotal.

However, marihuana does have a certain abuse potential (18, 98). In a follow-up study of ninety-seven regular users, 9 percent met operational criteria for abuse similar to criteria used in alcohol studies. In other words, marihuana use was associated with a variety of social, psychological, and medical problems in these individuals. Of the abusers, two-thirds had had traffic violations they attributed *solely* to marihuana intoxication; one-third had engaged in fights, again blaming marihuana. The latter finding was particularly interesting in view of the widespread observation that marihuana typically has antiaggressive effects. This, indeed, may be the case in most instances, but "set and setting" notoriously influence the effects of drugs, and it cannot be said any longer that marihuana has exclusively "taming" effects.

The difficulty of isolating marihuana effects is compounded by the fact that many regular users of marihuana also take other drugs. Furthermore, until the 1980s most marihuana available in this country contained small amounts of THC. Therefore, even though marihuana has been widely used in the United States, it is impossible to predict the consequences of prolonged use of high-potency cannabis by large numbers of people.

American attitudes toward marihuana have changed. Experimenters with the drug in the 1960s and 1970s commonly reported mild effects, probably because of the low potency of the drug mainly available at that time. As the THC content of available marihuana increased, there was a corresponding increase in the awareness of risks in using the drug. In 1986, for example, 71 percent of high school seniors believed that regular marihuana use involved risk, more than double the rate who believed this in 1978 (61). The change in attitude was accompanied by a modest decline in use. While the proportion of all Americans twelve and over who had ever used marihuana increased slightly (2 percent) from 1982 to 1985, those who had ever tried the drug in the younger groups declined, from 27 percent to 24 percent for twelve- to seventeen-year-olds, and from 60 percent to 44 percent for eighteen- to twenty-five-year-olds (60, 62).

Evidence accumulated over the past twenty years, suggests that regular use of marihuana is associated with the following hazards (19):

1. Bronchial and pulmonary irritation. Marihuana, because of its relatively poor combustibility, has up to 50 percent more polyaromatic hydrocarbons in its smoke than tobacco does (38). Cigarette for cigarette, the difference between tobacco and marihuana may be even more significant because of the way marihuana typically is smoked—down to a minuscule butt—and because the smoke itself is retained in the lung for a longer period than tobacco smoke. In one of the few human studies comparing adverse effects of cannabis and tobacco, measurements of bronchial constriction revealed that smoking less than one marihuana cigarette per day diminished vital capacity of the lungs as much as smoking sixteen tobacco cigarettes (87, 88).

2. Marihuana causes tachycardia. Smoking just one marihuana cigarette can significantly reduce exercise tolerance in heart patients with the anginal syndrome (4). This would militate against marihuana use by persons who have cardiac disorders.

3. Studies consistently show short-term adverse effects of marihuana on cognition and immediate memory (55).

4. Marihuana intoxication impairs reaction time, motor coordination, and visual perception (55). Studies under simulated conditions and in traffic have confirmed that the ability to operate a motor vehicle is adversely affected by marihuana use (48).

Immunoassays for cannabinoids have been available since the early 1980s. There are technical problems with the urine immunoassay currently being used and gas chromatography–mass spectrometry assays for TCH and its metabolites in blood and urine specimens are required for confirmation. One problem is that cannabinoids are detectable in the urine for at least one or two days after smoking and for weeks in some instances (49). For these reasons it has been difficult to estimate how many traffic accidents are associated with use of marihuana. Most marihuana users sometimes drive when they are high, according to an NIDA report (55).

References

1. *Adverse Reactions to Hallucinogenic Drugs*. Prepared by R. E. Meyer, p. 111. NIMH. Washington, D.C.: U.S. GPO, 1967.
2. Acute drug abuse reactions. Med. Lett. 27:77–80, 1985.
3. Andréasson, S., Allebeck, P., and Engström, A. Cannabis and schizophrenia. Lancet 2, 1987.
4. Aranow, W. S., and Cassidy, J. Effects of smoking marihuana and of a high-nicotine cigarette on angina pectoris. Clin. Pharmacol. Ther. 17: 549–554, 1975.
5. Aranow, R., and Done, A. Phencyclidine overdose: an emergency concept of management. J. Amer. Coll. Emerg. Physicians 7:56–59, 1978.
6. Bakker, C. B. The clinical picture in hallucinogen intoxication. Hosp. Med. 102–114, 1969.
7. Blachly, P. H. Management of the opiate abstinence syndrome. Amer. J. Psychiat. 122:742–744, 1966.
8. Bloomquist, E. R. *Marijuana*. Toronto, Canada: Glencoe Press, 1968.
9. Blum, R. H., *et al*. *The Utopiates: The Use and Users of LSD-25*. New York: Atherton Press, p. 304, 1964.
10. Brower, K. J., Hierholzer, R. and Maddahian, E. Recent trends in cocaine abuse in a VA psychiatric population. Hosp. Community Psychiat. 37:1229–1234, 1986.
11. Bucky, S. F. The relationship between background and extent of heroin use. Amer. J. Psychiat. 130:707–708, 1973.
12. Busto, U., Sellers, E. M., Naranjo, C. A., Cappell, H., Sanchez-Craig, M., and Sykora, K. Withdrawal reaction after long-term therapeutic use of benzodiazepines. N. Engl. J. Med. 315:854–859, 1986.
13. Childress, A. R., McLellan, A. T., and O'Brien, C. P. Abstinent opiate abusers exhibit conditioned craving, conditioned withdrawal and reductions in both through extinction. Brit. J. Addict. 81:655–660, 1986.
14. Chopra, G. S. Man and marijuana. Int. J. Addict. 4:215, 1969.
15. Climko, R. B., Roehrich, H., Sweeney, D. R., and Al-Razi, J. Ecstasy: a review of MDMA and MDA. Int. J. Psychiat. Med. 16:359–364, 1986.
16. Cohen, S. Angel dust. JAMA 238:515–516, 1977.

17. Cohen, S., and Edwards, A. E.   LSD and organic brain impairment. Drug Dependence, 1–4, 1968.
18. Committee to study the health-related effects of cannabis and its derivatives. National Academy of Sciences, Institute of Medicine. *Marijuana and Health*. Washington, D.C.: National Academy Press, 1982.
19. Council on Scientific Affairs. Marijuana, its health hazards and therapeutic potentials. JAMA 246:1823–1827, 1981.
20. Cregler, L. L., and Mark, H.   Medical complications of cocaine abuse. N. Engl. J. Med. 315:1495–1450, 1986.
21. Dependence on barbiturates and other sedative drugs. JAMA 193:107–111, 1965.
22. Dole, V. P., Nyswander, M. E., and Warner, A.   Successful treatment of 750 criminal addicts. JAMA 206:2708–2711, 1968.
23. Dole, V.   Research on methadone maintenance treatment. Proceedings of the 2d National Conference on Methadone Maintenance, 359–370, 1969.
24. Eddy, N. B., Halback, H., Isbell, H., and Seevers, M. H.   Drug dependence: its significance and characteristics. Bull. WHO 32:721–733, 1965.
25. Fleischhacker, W. W., Bamas, C., and Hackenberg, B.   Epidemiology of benzodiazepine dependence. Acta Psychiat. Scand. 74:80–85, 1986.
26. *Follow-up of Vietnam Drug Users* (Robins, L. N., principal investigator). National Institute of Drug Abuse: Washington. Special Action Office Monogr., Series A, no. 1, 1973.
27. Fraser, H. F., Wikler, A., Essig, C. F., and Isbell, H.   Degree of physical dependence induced by secobarbital or pentobarbital. JAMA 166:126–129, 1958.
28. Garey, R. E., Daul, G. C., Samuels, M. S., Ragan, F. A., and Hite, S. A. PCP abuse in New Orleans: a six-year study. Am. J. Drug Alcohol Abuse 13:135–140, 1987.
29. Garvey, M. J., and Tollefson, G. D.   Prevalence of misuse of prescribed benzodiazepines in patients with primary anxiety disorder or major depression. Am. J. Psychiat. 143:1601–1606, 1986.
30. Gawin, F. H.   New uses of antidepressants in cocaine abuse. Psychosomatics 27:24–29, 1986.
31. Ghodse, A. H.   Cannabis psychosis. Brit. J. Addict. 81:473–478, 1986.
32. Gold, M. S., and Galanter, M.   In *Cocaine: Pharmacology, Addiction and Therapy*, Stimmel, B. (ed.). New York: Haworth Press, 1987.
33. Goodman, L. S., and Gilman, A.   *The Pharmacological Basis of Therapeutics*. New York: Macmillan, 1965.
34. Goodwin, D. W.   Alcoholism. In *Clinical Psychopharmacology*, 1. Hippius, H., and Winokur, G. (eds.). Amsterdam: Excerpta Medica, 1983.
35. Goodwin, D. W.   *Anxiety*. New York: Ballantine Books, 1986.
36. Gossop, M., Bradley, B. and Phillips, G. T.   An investigation of withdrawal symptoms shown by opiate addicts during and subsequent to a 21-day inpatient methadone detoxification procedure. Addict. Behav. 12:1–5, 1987.

37. Grinspoon, L., and Bakalar, J. B. *Psychedelic Drugs Reconsidered*. New York: Basic Books, 1979.

38. Hoffman, D., Brunnemann, K. D., Gori, G. B., et al. Carcinogenicity of marihuana smoke. In *Recent Advances of Phytochemistry*, Vol. 9, Runeckles, V. C. (ed.). New York: Plenum, 1975.

39. Hollister, L. E. Drug tolerance, dependence and abuse. Kalamazoo, Mich.: Upjohn Co., 1985.

40. Hughes, P. H., et al. The social structure of a heroin coping community. Am. J. Psychiat. 128:551–558, 1971.

41. Hughes, P. H., Senay, E. C., and Parker, R. The medical management of a heroin epidemic. Arch. Gen. Psychiat. 27:585–593, 1972.

42. Inaba, D. S., Gay, G. R., Newmeyer, J. A., and Whitehead, C. Methaqualone abuse "luding out." JAMA 224:1505–1514, 1973.

43. Jaffe, J. H. Drug addiction and drug abuse. In *The Pharmacological Basis of Therapeutics*, Goodman, L. S., and Gilman, A. (eds.). New York: Macmillan, 1965.

44. Jain, N. C., Budd, R. D., and Budd, B. S. Growing abuse of phencyclidine: California "angel dust." N. Engl. J. Med. 297:673, 1977.

45. Johanson, C. E., and Uhlenhuth, E. H. Drug preferences in humans. Fed. Proc. 41:228–233, 1982.

46. Kieber, H. D., Topazian, M., Gaspari, J., Riordan, C. E., and Kosten, T. Clonidine and naltrexone in the outpatient treatment of heroin withdrawal. Am. J. Drug Alcohol Abuse 13:1–5, 1987.

47. Klepfisz, A., and Racy, J. Homicide and LSD, JAMA, 223:429–430, 1973.

48. Klonoff, H. Marihuana and driving in real-life situations. Science 186: 317–324, 1974.

49. Law, B., Pocock, K., and Moffat, A. C. An evaluation of homogeneous enzyme immunoassay (EMIT) for cannabinoid detection in biological fluids. J. Forensic Sci. Soc. 22:275–281, 1982.

50. Leighton, D. C., et al. *The Character of Danger: Psychiatric Systems in Selected Communities*, 3:322–353. New York: Basic Books, 1963.

51. Lerner, P. The precise determination of tetrahydrocannabinol in marihuana and hashish. Bull. Narc. 21:39–42, 1969.

52. Liden, C. B., Lovejoy, F. H., and Costello, C. E. Phencyclidine: nine cases of poisoning. JAMA 234:513–516, 1975.

53. Many "street" drugs are phony. Am. Druggist, 35–44, 1971.

54. *Marihuana and Health*, 2d Annual Report to Congress from the Secretary of HEW, 1972.

55. Marihuana and Health, 8th Annual Report to the U.S. Congress from the Secretary of HEW, HEW. Washington, D.C.: U.S. GPO 1980.

56. Mathias, D. W. Cocaine-associated myocardial ischemia. Am. J. Med. 81:675–678, 1986.

57. McAuliffe, W. E. A test of Wikler's theory of relapse: the frequency of relapse due to conditional withdrawal sickness. Int. J. Addict. 17:19–33, 1982.

58. McGlothlin, W. H., and Arnold, D. O. LSD revisited—a 10 year follow-up of medical LSD use. Arch. Gen. Psychiat. 24:35, 1971.
59. Mellman, T. A., and Uhde, T. W. Withdrawal syndrome with gradual tapering of alprazolam. Am. J. Psychiat. 143:1464–1469, 1986.
60. Millman, R. B., and Sbriglio, R. Patterns of use and psychopathology in chronic marijuana users. Psychiat. Clin. North Am. 9:533–538, 1986.
61. National High School Senior Survey. Rockville, Md.: NIDA, 1986.
62. National Household Survey on Drugs. Rockville, Md., NIDA, 1985.
63. Newman, R. G. Methadone treatment: defining and evaluating success. N. Engl. J. Med. 317:447–452, 1987.
64. Newman, R. G. Frustrations among professionals working in drug treatment programs. Brit. J. Addict. 82:115–120, 1987.
65. Patterns and Trends of Drug Abuse in the United States and Europe. Report to Congress by the NIDA. Rockville, Md., NIDA, 1987.
66. Paul, S. M., Hulihan-Giblin, B., and Skolnick, P. (+)-Amphetamine binding to rat hypothalamus: relation to anorexic potency of phenylethylamines. Science. 218:487–489, 1982.
67. Phencyclidine (PCP) abuse: an appraisal. NIDA Res. Monogr. 21. Rockville, Md.: NIDA, 1978.
68. Phillips, G. T., Gossop, M., and Bradley, B. The influence of psychological factors on the opiate withdrawal syndrome. Brit. J. Psychiat. 149: 235–240, 1986.
69. Position statement on psychoactive substance use and dependence: update on marijuana and cocaine. Am. J. Psychiat. 144:5, 1987.
70. Post, R. M. Cocaine psychoses: a continuum model. Am. J. Psychiat. 132:225–231, 1975.
71. Rainey, J. M., Jr., and Crowder, M. K. Prevalence of phencyclidine in street drug preparations. N. Engl. J. Med. 290:466–467, 1974.
72. Reed, D., Cravey, R. H., and Sedgwick, P. R. A fatal case involving phencyclidine. Int. Assn. Forensic Toxicologists Bull. 7:7, 1972.
73. Rickels, K., Case, W. G., Schweizer, E. E., Swenson, C., and Fridman, R. B. Low-dose dependence in chronic benzodiazepine users: a preliminary report on 119 patients. Psychopharmacol. Bull. 22:407-412, 1986.
74. Ritson, B., and Chick, J. Comparison of two benzodiazepines in the treatment of alcohol withdrawal: effects on symptoms and cognitive recovery. Drug Alcohol Depend. 18:329–334, 1986.
75. Rosser, W. W., Simms, J. G., Patten, D. W., and Forster, J. Improving benzodiazepine prescribing in family practice through review and education. Can. Med. Assoc. J. 124:147, 1981.
76. Roueché, B. Alcohol. New York: Grove Press, 1962.
77. Rubin, V., and Comitas, L. Ganja in Jamaica. The Hague and Paris: Mouton, 1975.
78. Scher, J. The marihuana habit. JAMA 214:1120, 1970.
79. Second Triennial Report to Congress on Drug Abuse and Drug Abuse Research. Rockville, Md.: NIDA, 1987.

80. Sharma, T. D.   Clinical observations of patients who used tetrahydro-cannabinol (THC) intravenously. Behav. Neuropsychiatry 4, 1972.
81. Showalter, C. V., and Thornton, W. E.   The increasing abuse of phen-cyclidine. Ill. Med. J. 151:387–389, 1977.
82. Siegel, R. K.   Cocaine hallucinations. Am. J. Psychiat. 135:3, 1978.
83. Siegel, R.   Cocaine freebase. J. Psychoactive Drugs, December 1982. (Special issue.)
84. Simon, E. J.   The opiate receptors. Annu. Rev. Pharmacol. Toxicol. 18:371–394, 1978.
85. Smart, R. G., and Bateman, K.   Unfavorable reactions to LSD: a review and analysis of available case reports. Can. Med. Assoc. J. 97:1214–1221, 1967.
86. Smith, D. E., and Wesson, D. R.   A new method for treatment of bar-biturate dependence. JAMA 213:294–295, 1970.
87. Tashkin, D. P., Shapiro, B. J., and Frank, I. M.   Acute effects of mari-huana on airway dynamics in spontaneous and experimentally induced bronchial asthma. Pharmacol. Marihuana 785–802, 1976.
88. Tashkin, D. P., Calvarese, B., and Simmons, M.   Respiratory status of 75 chronic marihuana smokers. Am. Rev. Respir. Dis. 117–261, 1978.
89. Terenius, L.   Endogenous peptides and analgesia. Annu. Rev. Pharma-colo. Toxicol. 18:189–204, 1978.
90. Tinklenberg, J. R., and Woodrow, K. M.   Drug use among youthful assaultive and sexual offenders. In Human Aggression, Frazier, S. H. (ed.), Baltimore: Williams & Wilkins, 1974.
91. Tjio, J., Pahnke, W. N., and Kurland, A. A.   LSD and chromosomes. JAMA 210:849–856, 1969.
92. Tong, T. G., Benowitz, N. L., Becker, C. E., et al.   Phencyclidine poisoning. JAMA 234:512–513, 1975.
93. Treatment of acute drug abuse reactions. Med. Lett. 29:83–86, 1987.
94. Vaillant, G. E.   A 12-year follow-up of New York narcotic addicts. Arch. Gen. Psychiat. 15:599, 1966.
95. Washton, A. M., and Gold, M. S.   Cocaine: A Clinician's Handbook. New York: Guilford Press, 1987.
96. Watson, R., Hartmann, E., and Schildkraut, J. J.   Amphetamine with-drawal: affective state, sleep patterns, and MHPG excretion. Am. J. Psychiat. 129:263–269, 1972.
97. Weinstein, S. P., Gottheil, E., Smith, R. H., and Migrala, K. A.   Co-caine users seen in medical practice. Am. J. Drug Alcohol Abuse 12:341–346, 1986.
98. Weller, R. A., and Halikas, J. A.   Objective criteria for the diagnosis of marihuana abuse. J. Nerv. Ment. Dis. 168:98–103, 1980.
99. Wetli, C. V.   Changing patterns of methaqualone abuse. JAMA 249:621–626, 1983.
100. Wikler, A.   Diagnosis and treatment of drug dependence of the bar-biturate type. Am. J. Psychiat. 125:6, 1968.
101. Winokur, A., Rickels, K., Greenblatt, D. J., Snyder, P. J., and Schatz,

N. J. Withdrawal reaction from long-term, low-dosage administration
of diazepam: a double-blind, placebo-controlled case study. Arch. Gen.
Psychiat. 37:101, 1980.

102. Zinberg, N. E., and Jacobson, R. C. The natural history of "chipping."
Am. J. Psychiat. 133:37–40, 1976.

9. Sociopathy (Antisocial Personality)

Definition

Sociopathy (sociopathic personality or antisocial personality) is a pattern of recurrent antisocial, delinquent, and criminal behavior that begins in childhood or early adolescence and is manifested by disturbances in many areas of life: family relations, schooling, work, military service, and marriage.

Historical Background

Prichard's 1835 monograph, A *Treatise on Insanity and Other Disorders Affecting the Mind,* is often cited as furnishing the first description of what is now called sociopathic personality (57). Under the label "moral insanity" he defined the disorder in this way: "The intellectual faculties appear to have sustained little or no injury, while the disorder is manifested principally or alone in the state of the feelings, temper, or habits. In cases of this description the moral and active principles of the mind are strongly perverted and depraved, the power of self government is lost or impaired and the individual is found to be incapable . . . of conducting himself with decency and propriety, having 'undergone a morbid change.'" As Craft makes clear (16), Prichard included under this rubric many examples of "temporary mental illness," and most of his cases were probably patients with affective disorder. Craft quotes Clouston (14): "Prichard quoted many such cases and vividly described the disease, but I should place most of his cases in my category of simple mania."

Clouston referred also to children "so constituted that they cannot be educated in morality on account of an innate brain deficiency . . . incapable of knowing . . . right and wrong. . . . Such moral idiots I, like others, have met with frequently . . . and persons with this want of development we may label under moral insanity."

Craft points out that Prichard used "moral" in three ways: first, in referring to "moral" treatment, meaning psychological treatment; second, in referring to emotional or affective responses in contrast to intellectual ones; and third, in an ethical sense of right or wrong. Most of the time, Prichard used the term in the first two ways and only incidentally in the last way.

As Craft also notes, Benjamin Rush in 1812 described "derangement of the moral faculties" as follows (62):

The moral faculty, conscience, and sense of deity are sometimes totally deranged. The Duke of Sully has given us a striking instance of this universal moral derangement in the character of a young man who belonged to his suite, of the name of Servin, who, after a life uncommonly distinguished by every possible vice, died, cursing and denying his God. Mr. Haslam has described two cases of it in the Bethlem Hospital, one of whom, a boy of 13 years of age, was perfectly sensible of his depravity, and often asked 'why God had not made him like other men.' . . . In the course of my life, I have been consulted in three cases of the total perversion of the moral faculties. One of them was in a young man, the second in a young woman, both of Virginia, and the third was in the daughter of a citizen of Philadelphia. The last was addicted to every kind of mischief. Her mischief and wickedness had no intervals while she was awake, except when she was kept busy in some steady and difficult employment. In all of these cases of innate, preternatural moral depravity, there is probably an original defective organization in those parts of the body which are occupied by the moral faculties of the mind.

It is Craft's conclusion that

Rush appears to give the first description of those with sound reason and good intellect who have an innate or lifelong irresponsibility without shame, being unchanged in affect, or by the consequences or by regard for others.

Thus Craft challenges the appropriateness of crediting Prichard with the first description.

Controversy over the concept of "moral insanity" developed partly in regard to the question whether the "morally insane" should be committed to mental hospitals or considered mentally ill in a court of law. The terms were gradually dropped as interest in the whole range of personality disorders grew.

In 1889 Koch (39) introduced the term "psychopathic inferiority" to imply a constitutional predisposition for many deviations of personality, including at least some that are now classified as anxiety disorders. Kraepelin (41), Kahn (37) and Schneider (65) proposed various classifications of personality disorders. Schneider's definition of psychopathic personalities as "all those abnormal personalities who suffer from their abnormalities or cause society to suffer" clearly included much more than "moral imbecility."

The term "psychopathic personality" was used inconsistently, sometimes to refer to the whole spectrum of deviant personalities and sometimes to a subgroup of antisocial or aggressive "psychopaths." Finally, to reduce confusion, the term "Sociopathic Personality Disturbance" was introduced for the latter group and was adopted in 1952 by the American Psychiatric Association's DSM-I. Nevertheless, many went on using the terms "psychopathy" and "sociopathy" interchangeably, whereas others continued to regard sociopathy as only one form of psychopathy. In a further attempt to reduce confusion, "Antisocial Personality" was adopted as the official diagnosis for the aggressive or antisocial psychopath or sociopath in later revisions of DSM and the International Classification of Disease. Sir Aubrey Lewis summarized the history of the concept (43). In recent years because of evidence in other countries that psychiatric diagnoses have been applied to political dissenters, some psychiatrists have been reluctant to use this diagnosis, despite the recognition that individuals with the clinical picture of antisocial personality are found all over the world and that only a few are political dissenters.

Epidemiology

Satisfactory data on the prevalence of antisocial personality are lacking. This is partly the result of the failure to reach a general agreement about a definition. It also stems from the difficulty of adequate ascertainment; we do not know what proportion of sociopaths comes to the attention of physicians. Nevertheless, sociopathy is seen frequently in psychiatric facilities, usually because of associated alcoholism and depression or because psychiatric care is made a condition of parole. In one series, 15 percent of male and 3 percent of female psychiatric outpatients were sociopaths (74).

Indirect estimates of population frequency, based on figures for juvenile delinquency and police trouble of all kinds, suggest that sociopathy is common, probably increasingly so, much more frequent in males than females, and in urban than rural environments, and most common in low socioeconomic groups (45, 60).

Sociopaths usually come from grossly disturbed families. Parental separation or divorce, early death, desertion, alcoholism, and criminality are characteristic. Only a small minority of sociopaths, in fact, come from families that are not characterized by one or more of these phenomena (15, 29, 36, 40).

Clinical Picture

Sociopathy begins in childhood or early adolescence (3, 52). The first manifestations may be those of the hyperactive child syndrome (51, 53, 63, 64, 73). Some investigators have demonstrated, however, that when hyperactivity is not accompanied by delinquent or antisocial behavior, it leads to antisocial personality patterns much less often. At the same time, delinquency associated with hyperactivity tends to be more severe than delinquency alone, with a worse prognosis for adult adjustment (55, 72). Restlessness, a short attention span, and unresponsiveness to discipline are common. Frequent fighting, often leading to conflicts with adults, and a history of being a general neighborhood nuisance are also common (11, 59).

A disturbed school history is characteristic of early sociopathy. In fact, a history of satisfactory school adjustment through high school is so unusual among sociopaths that either the diagnosis or the history should be questioned if it is present. Disruption of classes by talking out of turn, failure to pay attention to the teacher, fighting with classmates, arguing with the teacher, and even fighting with the teacher occur. Academic failures, truancy, and suspension often lead to school dropout or permanent expulsion (11, 30, 59).

Running way from home is also common, though it may be limited to a few one-night episodes. Occasionally, adolescent sociopaths will disappear for weeks or months. During such absences, they may wander over the country, hitchhiking, doing odd jobs, bumming around (59).

The sociopath's job history is characterized by poor performance. Undependability (being late, missing work, quitting without warning), inability to accept criticism and advice, frequent job changes without advancement, and being fired are typical (11, 59). Poor job performance coupled with limited and incomplete education result in low socioeconomic status, low income, and frequent need for financial assistance from family or society.

Sociopaths begin heterosexual experiences earlier than others and are much more likely to be promiscuous and indiscriminate in their sexual behavior. Homosexuality is probably more common among sociopaths than in the general population (12). Prostitution is frequent among female sociopaths, and homosexual prostitution is common among male sociopaths.

Early marriage is typical, especially among women. Very few sociopathic women, in fact, fail to marry. Their marriages are marked by infidelity, separation, and divorce. Male sociopaths have similar marital difficulties. Sociopaths tend to marry sociopaths; this is particularly true of women (12, 27).

Judges used to handle male delinquents by suspending punishment if the delinquent enlisted in the armed services. Generally, however, sociopaths do not do well in military service. Undependable and unable to conform to military discipline, they go AWOL, have difficulty with their superiors, are subjected to courtsmartial, and receive various nonhonorable discharges. The type of discharge is usually determined by the nature of the offenses, the philosophy of the commanding officer, and general military policy at the time.

Sooner or later most sociopaths have trouble with the police.

Some investigators, in fact, have required police trouble for the diagnosis of sociopathy (26). Stealing money from mother's purse or father's wallet or from a schoolmate may be an early sign of a budding criminal career. Shoplifting, peace disturbance (usually associated with drunkenness and fighting), various traffic offenses, auto theft, burglary, larceny, rape, robbery, and homicide may all occur. A sociopathic pattern of behavior is found in many convicted male felons (26, 28, 33) as well as in convicted female felons (12).

Many sociopaths engage in lying and the use of aliases, usually as understandable responses to social or legal difficulties, but sometimes without any obvious need to avoid punishment or retaliation. Such behavior has been called "pathological lying." Elaborate stories may be told to confuse or impress relatives. This behavior can take extreme forms such as masquerading as a physician, military officer, or businessperson. In time, relatives, friends, parole officers, and physicians learn to discount a good deal of what they are told by sociopaths.

Conversion symptoms (unexplained neurological symptoms) are common among sociopaths; in fact, sociopathy is second only to hysteria in producing such symptoms (31). Among sociopaths, conversion symptoms are characteristically associated with obvious social stresses such as police trouble. The full Briquet's syndrome (hysteria) may also be seen in association with sociopathy. Many female hysterics give a history of previous or concurrent sociopathy (12); many young delinquent and antisocial girls become hysterics as adults (12, 46, 59).

Most studies of delinquents and sociopaths indicate that their average IQ is below normal, but only a minority of delinquents and sociopaths suffer from significant mental deficiency, with an IQ less than 70 (16, 45). These studies have not all been adequately controlled for socioeconomic status, sibship size, and other variables correlated with intelligence. Low intelligence probably is not an important factor in the etiology of sociopathy.

Some studies have attempted to assess "constitutional" factors in sociopathy (68). Nonspecifically disordered electroencephalograms (EEGs) have been reported more frequently among sociopaths than controls, but many, if not most, sociopaths do not have abnormal EEGs (2, 23). Some authors have reported an association between delinquency, learning disability, and neurological impairment, but this has not been a completely consistent finding (71). Many studies have explored psychophysiological response patterns in sociopaths, and a theory of low autonomic and cortical arousal has been formulated to account for a persistent "stimulus hunger" or for the inability to learn socially approved behavior. The theory is thought to explain the sociopath's impulsive, excitement-seeking, and antisocial behavior (32, 50, 63). The research findings have not been entirely consistent (50), and some authors (49) have emphasized "a wider degree of variability in arousal levels and reactivity than [in] normal individuals." Reports have suggested an association between criminal behavior and chromosomal abnormalities. In particular, XYY karyotype, unusual height, and impulsive crimes have been associated, but not all studies are consistent (1, 19, 35, 54). On the other hand, more cases of the XXY karyotype (Klinefelter's syndrome) than of the XYY karyotype were found in a large Finnish series of male criminals referred for psychiatric evaluation (66). Studies of hormone levels, including androgens and adrenal steroids, have revealed inconsistent differences between sociopathic and nonsociopathic subjects (20, 42, 48).

A charming manner, lack of guilt or remorse, absence of anxiety, and a failure to learn by experience are said to be characteristic of sociopathy. The easygoing, open, and winning style, when present, presumably accounts for the sociopath's success as a confidence man. At best, however, the charm and appeal are superficial; often they are entirely absent.

When seen by psychiatrists, many sociopaths report anxiety symptoms, depression, and guilt (9, 74). These frequently occur with alcoholism. The guilt and remorse do not seem to lead to reduction of sociopathic behavior, however. This persistence of sociopathic behavior, despite repeated failure and punishment, is the basis for the statement that sociopaths "do not learn from experience." Suicide attempts are not rare, but completed suicide, in the absence of alcoholism or drug dependence, is infrequent (21).

Many people occasionally manifest antisocial or delinquent behavior. But only a minority demonstrate a consistent pattern of recurrent and repeated antisocial, delinquent, and criminal behavior beginning in childhood and lasting well into adulthood. The following case provides a vivid example of such persistent and pervasive behavior, justifying the diagnosis of antisocial personality.

A twenty-eight-year-old white man was seen in a hospital's emergency room because of repeated "spells" or "fits" while in jail; he had been arrested for driving while intoxicated, disturbing the peace, and resisting arrest. The patient's problem behavior began at about age seven or eight when his parents reported to the local school that they could not handle their son because he was "wild" and "couldn't sit still." Similar behavior had been noted by the school and the parents had been requested to meet with the teacher. From that time forward, a recurrent pattern became evident of fighting in school, other discipline troubles, poor academic performance, several periods of suspension, and several periods of truancy.

At the same time, the parents noted that the patient stole money from them and began to stay out late at night, presumably running around with a neighborhood gang. At about age fourteen, the boy had his first police contact: Several members of the gang were arrested for stealing from a number of neighborhood stores. Nothing came of this, but at age seventeen the patient was arrested for driving a stolen vehicle; he received a suspended sentence. Soon thereafter, he dropped out of school and started to work. His work record was most unsatisfactory, however, because of frequently missing days of work and because he was considered to be a troublemaker by his supervisors. As a result of this pattern, he failed at several jobs. He then enlisted in the army, where his inability to conform to expectations and accept military discipline resulted in a discharge "without honor," after about ten months.

Soon after leaving the army, he got married, but, as soon as his wife became pregnant, he deserted her and, at the same time, his parents became aware that he was drinking excessively on repeated occasions and that he was experimenting with a variety of illicit street drugs. He and his wife had a stormy marriage with frequent separations and reconciliations, though, after the birth of three children, they separated permanently. His wife accused him of physical abuse and repeated infidelity and also reported that he was physically aggressive toward the children, beating them without justification.

His work record continued to be erratic, and he was a frequent visitor to local medical clinics and emergency rooms with a variety of physical complaints. On one occasion, he was admitted briefly to a psychiatric unit because of taking an overdose of sleeping pills and some of the other drugs he was using.

In short, this patient's history is typical for antisocial personality. It includes all sorts of interpersonal and social difficulties, including problems and conflicts at school, with parents, on the job, in the army, with his wife and children, and with the police.

The diagnostic criteria from DSM-III-R are presented in Table 9.1.

Natural History

Sociopathy begins early in life (60). In some cases it may begin before the child starts attending school. Few sociopaths get through high school without recurrent difficulties. If the antisocial and delinquent pattern has not begun before age fifteen or sixteen, it is unlikely to occur. The prognosis may be better when the juvenile delinquency is "socialized," that is, when it occurs in a setting of close group involvement or loyalty (34). Lower social class, being female, and relative youth may adversely affect the clinical picture, but this is not a consistent finding (4).

The disorder is recurrent and varies in severity. Some sociopaths, generally milder cases, may remit during the late teens or early or mid-twenties (38). In other instances, sociopathic behavior persists into early middle age and then remits. Some sociopaths never

Table 9.1 Diagnostic criteria for Antisocial Personality Disorder (DCM-III-R)

A. Current age at least 18.
B. Evidence of Conduct Disorder with onset before age 15, as indicated by a history of *three* or more of the following:
   (1) was often truant
   (2) ran away from home overnight at least twice while living in parental or parental surrogate home (or once without returning)
   (3) often initiated physical fights
   (4) used a weapon in more than one fight
   (5) forced someone into sexual activity with him or her
   (6) was physically cruel to animals
   (7) was physically cruel to other people
   (8) deliberately destroyed others' property (other than by fire setting)
   (9) deliberately engaged in fire-setting
  (10) often lied (other than to avoid physical or sexual abuse)
  (11) has stolen without confrontation of a victim on more than one occasion (including forgery)
  (12) has stolen with confrontation of a victim (e.g., mugging, purse-snatching, extortion, armed robbery)
C. A pattern of irresponsible and antisocial behavior since the age of 15, as indicated by at least *four* of the following:
   (1) is unable to sustain consistent work behavior, as indicated by any of the following (including similar behavior in academic settings if the person is a student):
      (a) significant unemployment for six months or more within five years when expected to work and work was available
      (b) repeated absences from work unexplained by illness in self or family
      (c) abandonment of several jobs without realistic plans for others
   (2) fails to conform to social norms with respect to lawful behavior, as indicated by repeatedly performing antisocial acts that are grounds for arrest (whether arrested or not), e.g., destroying property, harassing others, stealing, pursuing an illegal occupation
   (3) is irritable and aggressive, as indicated by repeated physical fights or assaults (not required by one's job or to de-

fend someone or oneself), including spouse- or child-beating

(4) repeatedly fails to honor financial obligations, as indicated by defaulting on debts or failing to provide child support or support for other dependents on a regular basis

(5) fails to plan ahead, or is impulsive, as indicated by one or both of the following:

  (a) traveling from place to place without a prearranged job or clear goal for the period of travel or clear idea about when the travel will terminate

  (b) lack of a fixed address for a month or more

(6) has no regard for the truth, as indicated by repeated lying, use of aliases, or "conning" others for personal profit or pleasure

(7) is reckless regarding his or her own or others' personal safety, as indicated by driving while intoxicated, or recurrent speeding

(8) if a parent or guardian, lacks ability to function as a responsible parent, as indicated by one or more of the following:

  (a) malnutrition of child

  (b) child's illness resulting from lack of minimal hygiene

  (c) failure to obtain medical care for a seriously ill child

  (d) child's dependence on neighbors or nonresident relatives for food or shelter

  (e) failure to arrange for a caretaker for young child when parent is away from home

  (f) repeated squandering, on personal items, of money required for household necessities

(9) has never sustained a totally monogamous relationship for more than one year

(10) lacks remorse (feels justified in having hurt, mistreated, or stolen from another)

D. Occurrence of antisocial behavior not exclusively during the course of Schizophrenia or Manic Episodes.

---

improve. Attempts to explain remission have been based on hypothetical maturing or burning-out.

Remission, when it occurs, usually comes only after years of sociopathic behavior, during which education and work achievement have been severely compromised. Sociopaths rarely recover sufficiently to compensate for the "lost years." Thus "remission" usually means no more than a marginal social adjustment. Among many sociopaths, even if sociopathic behavior subsides, alcoholism and drug abuse persist and influence long-term adjustment (47).

Complications

The complications of sociopathy explain why the disorder is legitimately a medical concern. High rates of venereal disease, out-of-wedlock pregnancies, injuries from fights and accidents, alcoholism and drug dependence (8, 44), and various medical complications of these conditions mean that sociopaths often come to the attention of doctors. Furthermore, increased mortality from accidents and homicide contribute to a reduced life expectancy, particularly in early adulthood (26, 61).

Family Studies

As noted, sociopaths generally come from families that show severe social disturbances and disruption. Much of this family pathology consists of sociopathy and alcoholism (56).

In a study of male felons, most of whom were sociopaths, one-fifth of the first-degree male relatives were sociopaths and one-third were alcoholics (29). In a study of female felons, half of whom were sociopaths, one-third of the male relatives were sociopaths and one-half were alcoholics (13). In a study of children seen in a child guidance clinic (59), one-third of the fathers and one-tenth of the mothers of sociopaths were either sociopathic or alcoholic.

Among the attempts to determine whether this familial pattern reflects genetic factors have been investigations of twins and adoptees.

Twin studies generally have focused on antisocial behavior, delinquency, or criminality, without distinguishing between sociopaths and others who show these behavior patterns. Nevertheless, as most criminals apparently are sociopaths (26), this approach appears justified. In twin studies, the concordance rates with regard to behavior difficulties, delinquency, and criminality have nearly always been higher for monozygotic than dizygotic twins (50, 69). However, the differences have not been as great as those seen in other psychiatric illnesses where a genetic predisposition is suspected. Furthermore, all of the series have been small and the monozygotic–dizygotic difference in some does not reach statistical significance.

Twin studies of psychiatric illness may involve bias in ascertainment of cases because concordant cases may come to the attention of physicians more often than discordant cases. Population twin registries have been proposed as more suitable for unbiased ascertainment. Christiansen, having access to a register of all Danish twin births, was able to obtain a presumably unselected series of criminal twins (10). He found that monozygotic twins had a significantly higher concordance rate than dizygotic twins (36 percent vs. 12 percent).

Several follow-up studies of children of criminals and "psychopaths" indicate that these children, when adopted early in life by nonrelatives, are more likely to reveal psychopathic and criminal behavior as adults than are adopted children whose biological parents were not criminals or psychopaths (5, 6, 7, 17, 18, 67).

The association between sociopathy and hysteria within families and in the same individual (discussed in chap. 4) has been confirmed in other studies (70).

In summary, twin and adoptee studies at least suggest a genetic predisposition in some cases of sociopathy.

Differential Diagnosis

The differential diagnosis of sociopathy includes alcoholism, drug dependence, hysteria, schizophrenia, mania, and organic brain syndrome.

Alcoholism and drug abuse are frequent complications of sociopathy, and they aggravate antisocial and criminal patterns. In addition, however, some alcoholics and drug abusers who did not show sociopathic behavior in childhood or adolescence, before the onset of alcoholism or drug dependence, do show such behavior as a manifestation of the alcoholism or drug dependence. In these cases, the crucial diagnostic feature is the age of onset of antisocial and delinquent behavior; it is usually after age fifteen and coincident with, or following, the onset of the alcohol and drug abuse. When sociopathic behavior and alcohol or drug abuse all begin about the same time and before age fifteen, only follow-up may clarify the diagnosis, and often it is not possible to separate the sociopathy from the alcholism or drug dependence.

The familial and clinical associations between sociopathy and hysteria have already been noted. Many female sociopaths and an occasional male sociopath develop the picture of hysteria (30, 46). Moreover, there is an increased frequency of antisocial behavior and delinquency in the past histories and family histories of hysterics who are not sociopaths (30). Finally, hysteria and sociopathy cluster in the same families. All these observations suggest that the differential diagnosis between the two conditions may involve the recognition of overlapping manifestations of similar etiologic factors (25).

A small minority of male schizophrenics give a history of sociopathic patterns in adolescence, and a small number of young sociopaths, at follow-up, turn out to be schizophrenics (59), but the prevalence of antisocial personality is not increased in the biological relatives of schizophrenics compared to the relatives of controls (58). If evidence of schizophrenia has not appeared by the early twenties, however, it is unlikely to develop.

The behavior of some early onset manics may mimic that of sociopaths, especially in women. Occasionally it will be difficult to distinguish mania from sociopathy when the latter is complicated by amphetamine or other drug abuse. Follow-up should clarify the issue. Because mania before age fifteen is rare, a history of antisocial and delinquent behavior before that age would suggest sociopathy rather than mania.

Although antisocial and criminal behavior may accompany an organic brain syndrome, the latter rarely develops in childhood or early adolescence.

Clinical Management

A major problem in treating sociopathy is the patient's lack of motivation for change. Few sociopaths volunteer for treatment. They are nearly always brought to the physician's attention by pressure from schools, parents, or judges. Moreover, the commonly disturbed family situation and poor socioeconomic circumstances provide little support for any treatment program. Many therapists believe that early institutional therapy offers the only hope for success, but there is no consensus about this (24). Psychotherapy has not achieved impressive results (75).

It is particularly difficult to carry out treatment if associated alcoholism or drug dependence cannot be controlled. In many cases, remission of the alcoholism or drug abuse is accompanied by a reduction in antisocial and criminal behavior (26). Since sociopathy begins early and individuals at high risk can be readily recognized (22, 59), early case finding and treatment may ultimately offer hope of prevention.

References

1. Akesson, H. O., Forssman, H., Wahlström, J., and Wallin, L. Sex chromosome aneuploidy among men in three Swedish hospitals for the mentally retarded and maladjusted. Brit. J. Psychiat. 125:386–389, 1974.
2. Arthurs, R., and Cahoon, E. A clinical and electroencephalographic survey of psychopathic personality. Am. J. Psychiat. 120:875–877, 1964.
3. Behar, D., and Stewart, M. A. Aggressive conduct disorder of children. The clinical history and direct observations. Acta Psychiat. Scand. 65:210–220, 1982.
4. Behar, D., and Stewart, M. A. Aggressive conduct disorder: the influence of social class, sex and age on the clinical picture. J. Child. Psychol. Psychiat. 25:119–124, 1984.
5. Cadoret, R. J. Psychopathology in adopted-away offspring of biologic parents with antisocial behavior. Arch. Gen. Psychiat. 35:176–184, 1978.
6. Cadoret, R. J., Cunningham, L., Loftus, R., and Edwards, J. Studies of adoptees from psychiatrically disturbed biological parents. II. Temperament, hyperactive, antisocial and developmental variables. J. Pediatr. 87:301–306, 1975.
7. Cadoret, R. J., and Cain, C. Sex differences in predictors of antisocial behavior in adoptees. Arch. Gen. Psychiat. 37:1171–1175, 1980.

8. Cadoret, R. J., O'Gorman, T. W., Troughton, E., and Heywood, E. Alcoholism and antisocial personality. Arch. Gen. Psychiat. 42:161–167, 1985.
9. Chiles, J. A., Miller, M. L., and Cox, C. B. Depression in an adolescent delinquent population. Arch. Gen. Psychiat. 37:1179–1184, 1980.
10. Christiansen, K. O. Crime in a Danish twin population. Acta Genet. Med. Gemellol. 19:323–326, 1970.
11. Cleckley, H. *The Mask of Sanity.* St. Louis: C. V. Mosby, 1950.
12. Cloninger, C. R., and Guze, S. B. Psychiatric illness and female criminality: the role of sociopathy and hysteria in the antisocial woman. Am. J. Psychiat. 127:303–311, 1970.
13. Cloninger, C. R., and Guze, S. B. Psychiatric illness in the families of female criminals: a study of 288 first-degree relatives. Brit. J. Psychiat. 122:697–703, 1973.
14. Clouston, T. S. *Clinical Lectures on Mental Diseases.* London: Churchill, 1883.
15. Cowie, J., Cowie, V., and Slater, E. *Delinquency in Girls.* London: Humanities Press, 1968.
16. Craft, M. *Ten Studies into Psychopathic Personality.* Bristol: John Wright & Sons, 1965.
17. Crowe, R. R. The adopted offspring of women criminal offenders. Arch. Gen. Psychiat. 27:600–603, 1972.
18. Crowe, R. R. An adoption study of antisocial personality. Arch. Gen. Psychiat. 31:785–791, 1974.
19. Editorial. What becomes of the XYY male? Lancet 2:1297–1298, 1974.
20. Ehrenkranz, J., Bliss, E., and Sheard, M. H. Plasma testosterone: correlation with aggressive behavior and social dominance in man. Psychosom. Med. 36:469–475, 1974.
21. Garvey, M. J., and Spoden, F. Suicide attempts in antisocial personality disorder. Compr. Psychiat. 21:146–149, 1980.
22. Glueck, S., and Glueck, E. *Predicting Delinquency and Crime.* Cambridge,: Harvard Univ. Press, 1959.
23. Gottlieb, J. S., Ashby, M. C., and Knott, J. R. Primary behavior disorders and psychopathic personality. Arch. Neurol. Psychiat. 56:381–400, 1946.
24. Gralnick, A. Management of character disorders in a hospital setting. Am. J. Psychother. 33:54–66, 1979.
25. Guze, S. B. The role of follow-up studies: their contribution to diagnostic classification as applied to hysteria. Semin. Psychiat. 2:392–402, 1970.
26. Guze, S. B., Goodwin, D. W., and Crane, J. B. Criminality and psychiatric disorders. Arch. Gen. Psychiat. 20:583–591, 1969.
27. Guze, S. B., Goodwin, D. W., and Crane, J. B. A psychiatric study of the wives of convicted felons: an example of assortative mating. Am. J. Psychiat. 126:1773–1776, 1970.
28. Guze, S. B., Goodwin, D. W., and Crane, J. B. Criminal recidivism and psychiatric illness. Am. J. Psychiat. 127:832–835, 1970.

29. Guze, S. B., Wolfgram, E. D., McKinney, J. K., and Cantwell, D. P. Psychiatric illness in the families of convicted criminals. A study of 519 first-degree relatives. Dis. Nerv. Syst. 28:651–659, 1967.
30. Guze, S. B., Woodruff, R. A., Jr., and Clayton, P. J. Hysteria and antisocial behavior: further evidence of an association. Am. J. Psychiat. 127: 957–960, 1971.
31. Guze, S. B., Woodruff, R. A., and Clayton, P. J. A study of conversion symptoms in psychiatric outpatients. Am. J. Psychiat. 128:643–646, 1971.
32. Hare, R. D. Psychopathy: Theory and Research. New York: John Wiley & Sons, 1970.
33. Hare, R. D. Diagnosis of antisocial personality disorder in two prison populations. Am. J. Psychiat. 140:887–890, 1983.
34. Henn, F. A., Bardwell, R., and Jenkins, R. L. Juvenile delinquents revisited. Adult criminal activity. Arch. Gen. Psychiat. 37:1160–1163, 1980.
35. Hook, E. B. Behavioral implications of the human XYY genotype. Science 179:139–150, 1973.
36. Jonsson, G. Delinquent boys, their parents and grandparents. Acta Psychiat. Scand., Suppl. 195:43, 1967.
37. Kahn, E. Psychopathic Personalities. New Haven: Yale Univ. Press, 1931.
38. Kelso, J., and Stewart, M. A. Factors which predict the persistence of aggressive conduct disorder. J. Child Psychol. Psychiat. 27:77–86, 1986.
39. Koch, J. L. A. Leitfaden der Psychiatrie, 2d ed., Ravensburg: Dorn, 1889.
40. Koller, K. M., and Castanos, J. N. Family background in prison groups: a comparative study of parental deprivation. Brit. J. Psychiat. 117:371–380, 1970.
41. Kraepelin, E. Psychiatrie. Leipzig: J. A. Banth, 1909.
42. Kreuz, L. E., and Rose, R. M. Assessment of aggressive behavior and plasma testosterone in a young criminal population. Psychosom. Med. 34: 321–332, 1972.
43. Lewis, A. Psychopathic personality: a most elusive category. Psychol. Med. 4:133–140, 1974.
44. Lewis, C. E. Alcoholism, antisocial personality, narcotic addiction: an integrative approach. Psychiat. Dev. 3:223–235, 1984.
45. Lunden, W. A. Statistics on Delinquents and Delinquency. Springfield, Ill.: C. C. Thomas, 1964.
46. Maddocks, P. D. A five-year follow-up of untreated psychopaths. Brit. J. Psychiat. 116:511–515, 1970.
47. Martin, R. L., Cloninger, C. R., and Guze, S. B. The evaluation of diagnostic concordance in follow-up studies: II. A blind, prospective follow-up of female criminals. J. Psychiat. Res. 15:107–125, 1979.
48. Mattsson, A., Schalling, D., Olweus, D., Löw, H., and Svensson, J. Plasma testosterone, aggressive behavior, and personality dimensions in young male delinquents. J. Am. Acad. Child Psychiat. 19:476–490, 1980.
49. Mawson, A. R., and Mawson, C. D. Psychopathy and arousal: a new

interpretation of the psychophysiological literature. Biol. Psychiat. 12:49–74, 1977.
50. Mednick, S., and Christiansen, K. O. (eds.). *Biosocial Bases of Criminal Behavior.* New York: Gardner Press, 1977.
51. Mendelson, W., Johnson, N., and Stewart, M. A. Hyperactive children as teenagers: a follow-up study. J. Nerv. Ment. Dis. 153:273–279, 1971.
52. Mitchell, S., and Rosa, P. Boyhood behaviour problems as precursors of criminality: a fifteen-year follow-up study. J. Child Psychol. Psychiat. 22: 19–33, 1981.
53. Morrison, J. Adult psychiatric disorders in parents of hyperactive children. Am. J. Psychiat. 137:825–827, 1980.
54. Nielsen, J., and Henriksen, F. Incidence of chromosome aberrations among males in a Danish youth prison. Acta Psychiat. Scand. 48:87–102, 1972.
55. Offord, D. R., Sullivan, K., Allen, N., and Abrams, N. Delinquency and hyperactivity. J. Nerv. Ment. Dis. 167:734–741, 1979.
56. Oliver, J. E. Successive generations of child maltreatment. Brit. J. Psychiat. 147:484–490, 1985.
57. Prichard, J. C. *A Treatise on Insanity and Other Disorders Affecting the Mind.* London: Sherwood, Gilbert, & piper, 1835.
58. Rimmer, J., and Jacobsen, B. Antisocial personality in the biological relatives of schizophrenics. Compr. Psychiat. 21:258–262, 1980.
59. Robins, L. N. *Deviant Children Grown Up.* Baltimore: Williams & Wilkins, 1966.
60. Robins, L. N. The consequences of conduct disorder in girls. *Development of Antisocial and Prosocial Behavior.* New York: Academic Press, 1986.
61. Robins, L., and O'Neal, P. Mortality, mobility, and crime: problem children thirty years later. Am. Sociol. Rev. 23:162–171, 1958.
62. Rush, B. *Medical Inquiries and Observations upon the Diseases of the Mind.* New York: Hafner, 1962.
63. Satterfield, J. H. The hyperactive child syndrome: a precursor of adult psychopathy. Proceedings of the NATO Advanced Study Institute on Psychopathic Behavior. Les Arcs, France, September 1975. In *Psychopathic Behavior,* Hare, R. D. and Schalling, D. (eds.), 329–346. New York: John Wiley & Sons 1978.
64. Satterfield, J. H., Hoppe, C. M., and Schell, A. M. A prospective study of delinquency in 110 adolescent boys with attention deficit disorder and 88 normal adolescent boys. Am. J. Psychiat. 139:795–798, 1982.
65. Schneider, K. *Psychopathic Personalities.* London: Cassell, 1958.
66. Schröder, J., de la Chapelle, A., Hakola, P., and Virkkunen, M. The frequency of XYY and XXY men among criminal offenders. Acta Psychiat. Scand. 63:272–276, 1981.
67. Schulsinger, F. Psychopathy, heredity, and environment. Int. J. Ment. Health 1:190–206, 1972.
68. Shah, S. A., and Roth, L. H. Biological and psychophysiological factors

in criminality. In *Handbook of Criminology*, Glaser, D. (ed.), 101–173. Chicago: Rand McNally, 1974.

69. Slater, E., and Cowie, V. *The Genetics of Mental Disorders*. London: Oxford Univ. Press, 1971.

70. Spalt, L. Hysteria and antisocial personality. A single disorder? J. Nerv. Ment. Dis. 168:456–464, 1980.

71. Spreen, O. The relationship between learning disability, neurological impairment, and delinquency. Results of a follow-up study. J. Nerv. Ment. Dis. 169:791–799, 1981.

72. Stewart, M. A., Cummings, C., Singer, S., and deBlois, C. S. The overlap between hyperactive and unsocialized aggressive children. J. Child Psychol. Psychiat. 22:35–45, 1981.

73. Weiss, G., Hechtman, L., Milroy, T., and Perlman, T. Psychiatric status of hyperactives as adults: a controlled prospective 15-year follow-up of 63 hyperactive children. J. Am. Acad. Child Psychiat. 24:211–220, 1985.

74. Woodruff, R. A., Jr., Guze, S. B., and Clayton, P. J. The medical and psychiatric implications of antisocial personality (sociopathy). Dis. Nerv. Syst. 32:712–714, 1971.

75. Woody, G. E., McLellan, A. T., Luborsky, L., and O'Brien, C. P. Sociopathy and psychotherapy outcome. Arch. Gen. Psychiat. 42:1081–1086, 1985.

10. Brain Syndrome

Definition

Organic brain syndrome, or simply brain syndrome, is a clinical diagnosis based primarily on the mental status examination and usually applied to patients with recognizable medical and neurological disorders affecting brain structure and function. The diagnosis depends on finding impairment of orientation, memory, and other intellectual functions. Additional psychiatric symptoms may occur, including hallucinations, delusions, depression, obsessions, and personality change. Judgment is impaired. Patients may or may not be aware of having the disorder.

Brain syndromes may be divided into acute and chronic forms. An acute brain syndrome is brief and presumably reversible. *Delirium* is an acute brain syndrome associated with excitement and agitation, and often with hallucinations and delusions as well. A chronic brain syndrome, or *dementia*, is often progressive and chances of recovery are limited.

DSM-III-R distinguishes between organic *mental syndromes* and organic *mental disorders*. "Syndrome" is used when there is no reference to possible etiology. "Disorder" is used when etiology is known or presumed. Because both "syndrome" and "disorder" are based on mental-status findings, the distinction between the two terms seems unnecessary. The older term, "brain syndrome," will be preserved in this chapter for simplicity and to avoid semantic distinctions that probably do not contribute to the diagnosis and management of individual patients.

Historical Background

It is not known with any certainty when the clinical picture of brain syndrome was first recognized as distinct from other psychiatric disorders. On this subject, much interesting material is presented in an annotated anthology of selected medical texts (42).

One of the earliest descriptions in English was provided in 1615 by a clergyman, Thomas Adams, in his treatise *Mystical Bedlam, the World of Mad-Men* (1). He referred to "some *mad*, that can rightly judge of the things they see, as touching *imagination and phantasie:* but for *cogitation* and *reason*, they swarve from naturale judgment."

In 1694 William Salmon (89) described in detail a case of dementia and noted that the patient was "not mad, or distracted like a man in Bedlam . . . (but) decayed in his Intellectuals."

In 1761 Giovanni Battista Morgagni, the great Italian pioneer in morphologic pathology, who first systematically correlated clinical features and course with postmortem findings, described certain areas of "hardness" in the brains of former mental patients (74). He emphasized, however, that the correlation between clinical picture and anatomical findings was not consistent.

Morgagni's work set the stage for Antoine Laurant Jessé Bayle who, in 1822, published the first systematic clinicopathologic study of paresis. In this study the clinical symptoms, including progressive dementia, were correlated with changes in brain parenchyma and meninges (10). Bayle's findings were confirmed in 1826 by Louis Florentin Calmeil (22).

Another milestone in the understanding of brain syndrome was Korsakoff's investigation of the disorder that bears his name (103). His observations, reported between 1887 and 1891, correlated a particular form of dementia manifested by extreme amnesia with lesions of the brainstem.

Over the years the accumulation of case reports and clinical experience indicated that disorientation and memory impairment were the hallmarks of a clinical syndrome seen frequently in the presence of demonstrable brain damage, systemic infections, or intoxications. But systematic investigations were infrequent. Two modern studies based on autopsy results indicate that brain syndromes are associated with brain pathology in the majority of cases (25, 48).

Epidemiology

There have been few epidemiologic studies of the actue brain syndrome. It occurs commonly in the presence of pneumonia, systemic infections, congestive heart failure, high fever, fluid and electrolyte imbalance, stroke, the postoperative state, and intoxication with alcohol or other drugs, but the exact frequency is unknown. Reports have dealt with cases referred for psychiatric consultation because of disturbed behavior rather than with the frequency of the syndrome as determined by systematic study.

It is known that dementia of the Alzheimer's type and vascular disease of the brain are two of the most common causes of chronic brain syndromes and that their prevalence increases with age. But epidemiologic studies of chronic brain syndrome have serious flaws. Most have been based on hospitalization rates. Different rates have been reported related to urban versus rural status, sex, race, educational achievement, and marital status. Many authors, however, emphasize the need for caution in extrapolating from hospitalization rates to true prevalence (32, 67, 68). A review of recent field surveys indicates inconsistent findings in regard to these same social and economic factors (32), making it clear that most such correlations can only be regarded as tentative.

Clinical Picture

The acute brain syndrome is nearly always associated with medical, surgical, or neurological disorders, or with drug intoxications. This association is so consistent that an unexplained acute brain syndrome should alert the physician to the likelihood that such a disorder exists or is developing. A patient with pneumonia, for example, will sometimes present with an acute brain syndrome hours before the other clinical findings appear.

An acute brain syndrome may vary clinically, but two features are of special diagnostic importance: disorientation and impairment of recent memory. Without at least one of these, the diagnosis can only be suspected.

Patients may complain that they do not understand where they are and that they cannot find the bathroom; or they may bitterly insist that no one has fed them, even though the remains of the last meal are on the tray next to their bed. Other common clinical features include a depressed or fearful mood, apathy, irritability, impaired judgment, suspiciousness, delusions, hallucinations, and combative, uncooperative, or frightened behavior.

A labile, shifting affect with sudden crying spells; an argumentative, demanding manner ("I've called and called and no one pays any attention to me"); visual hallucinations of frightening or threatening faces ("They're grinning at me all the time, . . . they know something bad will happen"); unreasonable pressure for changes in room, personnel, or food; inappropriate efforts to get dressed and leave the hospital or to "escape" other patients who "want to hurt me" are examples of such clinical features.

The combination of disorientation, impaired memory, suspiciousness, hallucinations, and combative or frightened behavior constitutes delirium. This is the most dramatic, clinically severe form of the acute brain syndrome. It may develop rapidly without any preceding manifestations or gradually in a patient who has been quietly confused and apathetic for many hours or days. Sometimes it may complicate a chronic state of dementia. The diagnostic criteria for delirium, according to DSM-III-R, are presented in Table 10.1.

An acute brain syndrome may fluctuate strikingly from day to day and even from hour to hour. Some patients show a diurnal pattern with the most obvious and severe manifestations at night. They may appear normal in the morning except for haziness about the night before and yet be frankly disoriented, confused, and hallucinating that night. In some cases, the patient's mental status must be checked repeatedly to elicit disorientation or recent memory difficulty.

Disorientation may relate to time, place, or person. The first is the most common, the last the least common. Minor errors by patients who have been sick or hospitalized for some time should not receive undue emphasis. But persistent inability to recall the month or year correctly, especially if the correct answers have been

Table 10.1 Diagnostic criteria for Delirium (DSM-III-R)

A. Reduced ability to maintain attention to external stimuli (e.g., questions must be repeated because attention wanders) and to appropriately shift attention to new external stimuli (e.g., perseverates answer to a previous question).

B. Disorganized thinking, as indicated by rambling, irrelevant, or incoherent speech.

C. At least two of the following:
    (1) reduced level of consciousness, e.g., difficulty keeping awake during examination
    (2) perceptual disturbances: misinterpretations, illusions, or hallucinations
    (3) disturbance of sleep-wake cycle with insomnia or daytime sleepiness
    (4) increased or decreased psychomotor activity
    (5) disorientation to time, place, or person
    (6) memory impairment, e.g., inability to learn new material, such as the names of several unrelated objects after five minutes, or to remember past events, such as history of current episode of illness

D. Clinical features develop over a short period of time (usually hours to days) and tend to fluctuate over the course of a day.

E. Either (1) or (2):
    (1) evidence from the history, physical examination, or laboratory tests of a specific organic factor (or factors) judged to be etiologically related to the disturbance
    (2) in the absence of such evidence, an etiologic organic factor can be presumed if the disturbance cannot be accounted for by any nonorganic mental disorder, e.g., Manic Episode accounting for agitation and sleep disturbance

---

offered recently, recurrent failure to identify where the patient is, or repeated inability to recall recent events correctly, are diagnostic. Because the manifestations are often more definite at night, the first indication of an acute brain syndrome may be recorded in the nurse's notes. A report of the patient mistaking the nurse for someone else—such as a neighbor or relative ("What is my sister doing here?") or experiencing hallucinations—is sometimes ignored until the patient is frankly disoriented and combative.

An acute brain syndrome seen in a general hospital usually indicates that the patient is seriously ill physically. Available evidence suggests that when the condition interferes with nursing care or treatment or when it disturbs other patients, the prognosis is poorer than is the case for patients with similar medical illnesses matched for age, sex, and race and who do not have an acute brain syndrome (48, 49, 81).

Diagnostic criteria for dementia, according to DSM-III-R, are presented in Table 10.2. Many patients with dementia may present

Table 10.2 Diagnostic criteria for Dementia (DSM-III-R)

A. Demonstrable evidence of impairment in short- and long-term memory. Impairment in short-term memory (inability to learn new information) may be indicated by inability to remember three objects after five minutes. Long-term memory impairment (inability to remember information that was known in the past) may be indicated by inability to remember past personal information (e.g., what happened yesterday, birthplace, occupation) or facts of common knowledge (e.g., past Presidents, well-known dates).

B. At least one of the following:
    (1) impairment in abstract thinking, as indicated by inability to find similarities and differences between related words, difficulty in defining words and concepts, and other similar tasks
    (2) impaired judgment, as indicated by inability to make reasonable plans to deal with interpersonal, family, and job-related problems and issues
    (3) other disturbances of higher cortical function, such as aphasia (disorder of language), apraxia (inability to carry out motor activities despite intact comprehension and motor function), agnosia (failure to recognize or identify objects despite intact sensory function), and "constructional difficulty" (e.g., inability to copy three-dimensional figures, assemble blocks, or arrange sticks in specific designs)
    (4) personality change, i.e., alteration or accentuation of premorbid traits

C. The disturbance in A and B significantly interferes with work or usual social activities or relationships with others.

D. Not occurring exclusively during the course of Delirium.

E. Either (1) or (2):
   (1) there is evidence from the history, physical examination, or
        laboratory tests of a specific organic factor (or factors)
        judged to be etiologically related to the disturbance
   (2) in the absence of such evidence, an etiologic organic factor
        can be presumed if the disturbance cannot be accounted
        for by any nonorganic mental disorder, e.g., Major Depres-
        sion accounting for cognitive impairment

Criteria for severity of Dementia:

Mild: Although work or social activities are significantly impaired,
the capacity for independent living remains, with adequate personal hygiene and relatively intact judgment.

Moderate: Independent living is hazardous, and some degree of
supervision is necessary.

Severe: Activities of daily living are so impaired that continual
supervision in required, e.g., unable to maintan minimal personal
hygiene; largely incoherent or mute.

---

with depressive symptoms (63) or somatic complaints such as
headache, abdominal pain, and constipation. Others are brought
to physicians by relatives because of temper outbursts, socially embarrassing behavior, or suspiciousnes. The cognitive and memory
impairment may become evident only as the patient's history is
elicited. DSM-III-R provides additional diagnostic criteria for subtypes with prominent delusions or prominent disturbances of
mood; it remains to be seen whether such subdivisions will prove
useful.

Depressed patients may complain of poor memory and some
studies have revealed impairment of short-term memory in many.
The memory difficulty improves with remission of the depression
(95). Sometimes, however, it is important to distinguish between
complaints of memory difficulty (pseudodementia) and *actual* impairment of memory. In some middle-aged and older patients suffering from depression, the complaint of memory impairment is
disproportionately greater than the degree of memory impairment

noted on systematic testing, and the perceived memory difficulty subsides as the depression improves (23, 51, 60, 71, 107).

In dementia, the patient's insight may vary greatly. Sometimes, especially earlier in the illness, the patient may volunteer that, "I keep losing things. I don't know where I left them. I occasionally can't remember the directions when I drive home." Often, however, and especially as the illness progresses, the patient may deny any memory difficulty and insist that he was preoccupied or distracted by worries or interruptions. Inability to follow directions and confusion about what others intend may lead to irritability and anger: "Nobody tries to explain things to me. Nobody cares about how I feel—I'll show them." Frustration at being unable to accomplish specific tasks, including specific cognitive tests, frequently leads to suspicion and resistance: "That's a silly thing to do. I could do it if I really wanted to, but I don't see a good reason for it. You're trying to trip me up and fool me."

Inattention to cleanliness and grooming, deterioration in table manners, the use of abusive and foul language, social withdrawal and a generally inconsiderate manner, sudden and seemingly unprovoked outbursts of anger and even physical violence present increasingly heavy burdens to the family as the disorder worsens and may require institutionalization.

The most important point about dementia is that in some cases the underlying illness is treatable (27, 51, 104, 105). These cases should be recognized as early as possible because recovery may be related to the duration of the dementia. Drug intoxication (12, 79, 91, 101, 102), liver failure (90), hypothyroidism (36, 94), pernicious anemia (97), paresis (43), subdural hematomas and benign brain tumors (6), and normal pressure hydrocephalus (59, 64, 83) are infrequent but recognizable causes of dementia, though debate continues about the diagnosis and validity of the last entity (4). Recurrent hypoglycemia because of insulin treatment of diabetes, pancreatic islet cell tumors, or the too rapid alimentary absorption of glucose in patients who have undergone subtotal gastric resection may also lead to severe dementia (7, 8, 20). All these disorders may be treated. Chronic alcohol abuse has been associated with a memory impairment (11, 65, 85) that is distinct from the Wernicke-Korsakoff syndrome and that, if recognized early, often can be reversed by abstention from further drinking; however, debate continues about the validity of an alcohol-induced dementia separate from Wernicke's encephalopathy or hepatic encephalopathy (99). Recent work has suggested that the dementia frequently associated with chronic renal dialysis may be caused by aluminum intoxication from the water used in such great quantities during dialysis (3, 5, 72, 82). Although probably not readily reversible, this syndrome may now be preventable.

Most cases of dementia, however, are the result of intrinsic brain disease or arteriosclerotic changes in the blood vessels supplying the brain (104). Studies by Roth (88) suggest that in both groups of patients there is "a fairly strict quantitative relationship . . . between the amount of intellectual deterioration . . . and the amount of cerebral damage at postmortem. In each group of disorders there are . . . threshold effects, suggesting that up to a certain limit the destruction wrought by degenerative changes can be accommodated within the reserve capacity of the brain." In arteriosclerotic dementia, the intellectual deterioration is related to the volume of brain softening; in demnetia of the Alzheimer's type, to the density of senile plaques. Recent work with positron-emission tomography (PET) indicates that, in both conditions, the severity of dementia is correlated with a decline in cerebral blood flow and mean cerebral oxygen utilization (41).

Recent usage indicates a preference for the term multi-infarct dementia rather than arteriosclerotic dementia. This change reflects the recognition that the dementia results from loss of brain tissue, typically with a deterioration in intellectual function proportionate to the amount lost.

Other causes of dementia, still untreatable, are a variety of presenile brain degenerations and traumatic injuries to the brain (45, 57, 61, 69).

For many years a distinction was made between Alzheimer's disease, a presenile dementia, and senile brain disease or dementia, despite the fact that the anatomical findings (senile plaques, neurofibrillary tangles, and granulovacuolar degenerative changes) in the two groups of patients are similar. In current practice the two conditions are viewed as identical (86, 106).

Recent work indicates that dementia of the Alzheimer's type, whether preseniile or senile, is associated with loss of central cholinergic neurons in the amygdala, hippocampus, and cortex (24, 30, 37, 87, 92, 100). Of special interest is evidence that certain cholinergic nuclei located in the basal forebrain and contiguous areas (particularly the *nucleus basalis* of Meynert) are largely destroyed in such patients (80, 109). Other biochemical abnormalities involving brain catecholamines have also been described (2, 26). Finally, recent work strongly suggests that the neurofibrillary tangles characteristic of dementia of the Alzheimer's type originate from neurotubules (46).

Of great interest is the recent findings that the genetic defect in familial Alzheimer's disease is probably located on chromosome 21, the same chromosome that is involved in Down's syndrome (44, 98). This finding is of special interest, in view of the observation that changes similar to those seen in Alzheimer's disease are also found in patients with Down's syndrome who survive until early middle age (28, 55).

Related to the findings concerning the role of impaired cholinergic function in dementia of the Alzheimer's type is work suggesting that age-related memory disturbances in animals and humans may also be associated with cholinergic dysfunction (9, 31, 77). Whether the problem in Alzheimer patients is simply a more rapid progression of age-related changes or whether it involves certain more specific factors is not yet clear.

Huntington's disease, another form of severe dementia, long recognized as caused by a dominant gene, is now clearly linked to chromosome 4 (47, 70). At the same time, several investigators have emphasized an association between Huntington's disease and affective disorder, noting that the affective disorder often may be an early manifestation of the Huntington's disease (21, 39, 40). This association may help account for clinical observations that suggest suicide is a frequent complication of Huntington's disease, even when little or no dementia is present. The possibility of testing presymptomatically for the disorder by using a deoxyribonucleic acid (DNA) marker raises obvious and difficult ethical and legal concerns (13) about the implications involved in testing for a serious illness for which there is no known treatment or prevention.

Another form of dementia recently has been associated with the central nervous system (CNS) involvement in acquired immune deficiency syndrome (AIDS). Indeed, dementia may be an early or even the first clinical manifestation of human immunodeficiency virus (HIV) infection (56, 76, 93, 96).

The use of computerized tomography (CT) may contribute valuable information for the diagnosis of dementia, but it is important to remember that the correlation between cognitive impairment and the radiological findings is sometimes poor. It is likely, however, that refinements in CT will improve such correlations in the future (17).

Natural History

As noted, an acute brain syndrome may occur in the course of various medical illnesses. Often it is impossible to determine the most crucial factor in a patient with simultaneous heart failure, infection, fever, and dehydration who is also receiving a variety of medications. Generally, the brain syndrome subsides as the underlying abnormalities are corrected. Sometimes, when patients have been very sick and have had a prolonged brain syndrome, many days pass after medical abnormalities are controlled before the mental picture clears. Acute brain syndromes resulting from drug intoxication or drug withdrawal (e.g., alcohol or barbiturates) nearly always subside within a few days after discontinuing the drug.

Chronic brain syndromes usually develop insidiously. The early manifestations may be subtle and only in retrospect does their significance become evident. Fatigability, moodiness, distractibility, depression, irritability, and carelessness may be present long before memory difficulty, intellectual deterioration, and disorientation can be clearly and easily detected. Depending on the underlying brain disease, the dementia may either progress to total incapacity and death or stabilize for long periods (16, 51, 66, 86, 106). As in acute brain syndrome associated with delirium, chronic brain syndrome also appears to be associated with an excess mortality (14).

Complications

Faulty judgment in important decisions, inability to care for oneself, accidents, and suicide are the principal complications of brain syndromes.

One of the more difficult decisions physicians must make concerns patients with chronic brain syndromes whose behavior raises questions about their mental competence. At some point in the course of many chronic brain syndromes, the patients' confusion, forgetfulness, temper outbursts, and questionable financial dealings will lead relatives and friends to question whether they are still competent to handle their affairs and care for themselves properly. Losing or giving away large sums of money, writing bad checks, extreme carelessness of dress, unprecedented sexual behavior (e.g., molesting children or displaying genitals), wandering away from home and getting lost, and unpredictable temper displays may require legal action to protect patients from their own acts and to permit extended institutionalization.

Chronic brain syndrome, complicated by secondary depression, is one of the psychiatric disorders associated with suicide risk, though it accounts for only a small portion of suicides (84).

In some cases of delirium it is not possible to tell whether the patient jumped from a window with the intention of committing suicide or fell because of confusion and fear.

The role of dementia—especially that of Alzheimer's disease—as an important risk factor for serious falls has only recently gained recognition (75).

Family Studies

Investigations of the familial prevalence of organic brain syndromes have been concerned primarily with specific disorders that lead to dementia such as Huntington's chorea and senile brain disease. As noted earlier, the genetic basis of Huntington's chorea, which is carried by an autosomal dominant gene, is established (78), and senile dementia is apparently increased, perhaps fourfold, among the relatives of index cases (18, 19, 62, 73). Similarly, the familial, often genetic, basis of certain rare neurological disorders leading to dementia is clear (78). Yet there is no evidence of any general familial predisposition to brain syndromes.

Of considerable theoretical interest, however, is the observation that Alzheimer's disease, Down's syndrome, and myeloproliferative disorders cluster in the same families. Coupled with the finding that changes in the brain characteristic of Alzeimer's disease and associated with a dementing illness are seen in most persons with trisomy 21 (Down's syndrome) who die as they approach middle age, this familial pattern has led to the suggestion that an important common denominator may be a genetically determined defect in neuronal microtubules (29, 52, 53, 54), though another study failed to find such association (108).

Recent evidence (15) suggests that individuals suffering from the Wernicke-Korsakoff syndrome may have a genetic predisposition to thiamine deficiency that becomes clinically significant when they are exposed to a poor diet, a possible explanation of why only some chronic alcoholics develop the syndrome.

Differential Diagnosis

Other psychiatric conditions may be mimicked by brain syndromes: patients' anxiety attacks may suggest anxiety neurosis; low mood and apathy, an affective disorder; hallucinations and delusions, schizophrenia. The crucial question in each case is whether the patient exhibits definite disorientation or memory impairment. These mental status abnormalities are pathognomonic of a brain syndrome; they are not manifestations of an uncomplicated, so-called functional disorder. If a patient with another psychiatric illness develops disorientation or memory impairment, one should suspect that something else has developed: a drug reaction or medical or neurologic illness.

When a patient is unable or unwilling to cooperate in the mental status examination, an acute brain syndrome may be suspected. If there has been a sudden change in his behavior, speech, or manner and if the behavioral change develops in a clinical situation that frequently predisposes to an acute brain syndrome, the diagnosis of brain syndrome should be considered. A definitive diagnosis must await evidence of disorientation or memory impairment; these will usually become apparent as the patient is observed carefully.

Clinical Management

Correction of the underlying medical or neurological condition, whenever possible, is the principal aim of therapy for patients with brain syndromes. At the same time, certain measures often help in the management of the brain syndrome itself. Good nursing care is important. A calm, sympathetic, reassuring approach can turn a frightened, combative patient into a quiet, cooperative one. A patient with an acute brain syndrome often misinterprets stimuli and has unpredictable emotional responses. It is important, therefore, to provide a familiar, stable, unambiguous environment for such patients. Repeated simple explanations and frequent reassurance from familiar nurses, attendants, or relatives may be helpful. Patients do better with constant light; shadows or the dark easily frighten them.

Only the smallest effective doses of drugs acting on the CNS should be used, because patients with a brain syndrome are frequently sensitive to these agents. Brain syndromes are, in fact, often precipitated by sedatives or hypnotics and may subside when such drugs are discontinued. No drug is entirely safe. Careful attention to dosage and mental status are more important than the particular sedative or hypnotic used.

Some patients need to be kept from harming themselves. Usually, a relative, friend, or attendant who is able to be with the patient constantly—talking to him, explaining things, reassuring him—can calm him enough to permit appropriate care without restraints on a general hospital service. When this does not work or is not possible, rather than risk having a confused and frightened patient fall out of bed or jump out a window, transfer to the psychiatric service or physical restraint may be necessary.

If a psychiatric unit is not available or if the patient's medical condition and treatment require bed rest, a body restraint may be necessary. Obviously, a restrained patient should be watched carefully, and the previously described measures should be continued with the hope of calming him quickly, making any period of restraint brief.

Organized groups for relatives of demented patients can be very helpful in providing emotional support and practical suggestions for handling the wide range of problems that dementia presents to families (42). Too often the impact on family members is overwhelming, and this must be considered by the physician so that appropriate steps can be taken to reduce the burden, at least intermittently (33, 34, 35). Similarly, the introduction of systematic genetic counseling and family support for relatives of patients with Huntington's disease can offer much needed assistance to those who are at high risk for developing the disorder (50).

References

1. Adams, T. *Mystical Bedlam, the World of Mad-Men.* London, 1615.
2. Adolfsson, R., Gottfries, C. G., Roos, B. E., and Winblad, B.   Changes In the brain catecholamines in patients with dementia of Alzheimer type. Brit. J. Psychiat. 135:216–223, 1979.
3. Alfrey, A. C., LeGendre, G. R., and Kachny, W. D.   The dialysis encephalopathy syndrome. N. Engl. J. Med. 294:184–188, 1976.
4. Anderson, M.   Normal pressure hydrocephalus. Brit. Med. J. 293:837–838, 1986.

5. Arieff, A. I., Cooper, J. D., Armstrong, D., and Lazarowitz, V. C. Dementia, renal failure, and brain aluminum. Ann. Intern. Med. 90:741–747, 1979.
6. Avery, T. L. Seven cases of frontal tumour with psychiatric presentation. Brit. J. Psychiat. 119:19–23, 1971.
7. Bale, R. N. Brain damage in diabetes mellitus. Brit. J. Psychiat. 122:337–342, 1973.
8. Banerji, N. K., and Hurwitz, L. J. Nervous system manifestations after gastric surgery. Acta Neurol. Scand. 47:485–513, 1971.
9. Bartus, R. T., Dean, R. L., III, Beer, B., and Lippa, A. S. The cholinergic hypothesis of geriatric memory dysfunction. Science 217:408–417, 1982.
10. Bayle, A. L. J. Recherches sur l'arachnitis chronique. M.D. Thesis, Paris, 1822.
11. Berglund, M., Gustafson, L., Hagberg, B., Ingvar, D. H., Nilsson, L., Risberg, J., and Sonesson, B. Cerebral dysfunction in alcoholism and presenile dementia. Acta Psychiat. Scand. 55:391–398, 1977.
12. Bergman, H., Borg, S., and Holm, L. Neuropsychological impairment and exclusive abuse of sedatives or hypnotics. Amer. J. Psychiat. 137:215–217, 1980.
13. Bird, S. J. Presymptomatic testing for Huntington's disease. JAMA 253:3286–3291, 1985.
14. Black, D. W., Warrack, G., and Winokur, G. The Iowa record-linkage study. II. Excess mortality among patients with organic mental disorders. Arch. Gen. Psychiat. 42:78–81, 1985.
15. Blass, J. P., and Gibson, G. E. Abnormality of a thiamine-requiring enzyme in patients with Wernicke-Korsakoff syndrome. N. Engl. J. Med. 297:1367–1370, 1977.
16. Blessed, G., and Wilson, I. D. The contemporary natural history of mental disorder in old age. Brit. J. Psychiat. 141:59–67, 1982.
17. Bondareff, W., Baldy, R., and Levy, R. Quantitative computed tomography in senile dementia. Arch. Gen. Psychiat. 38:1365–1368, 1981.
18. Breitner, J. C. S., Folstein, M. F., and Murphy, E. A. Familial aggregation in Alzheimer dementia—I. A model for the age-dependent expression of an autosomal dominant gene. J. Psychiat. Res. 20:31–43, 1986.
19. Breitner, J. C. S., Murphy, E. A., and Folstein, M. F. Familial aggregation in Alzheimer dementia—II. Clinical genetic implications of age-dependent onset. J. Psychiat. Res. 20:45–55, 1986.
20. Burton, R. A., and Raskin, N. H. Alimentary (post gastrectomy) hypoglycemia. Arch. Neurol. 23:14–17, 1970.
21. Caine, E. D., and Shoulson, I. Psychiatric syndromes in Huntington's disease. Am. J. Psychiat. 140:728–733, 1983.
22. Calmeil, L. F. De la paralysie considérée chez les aliénés. Paris, 1826.
23. Cavenar, J. O., Jr., Maltbie, A. A., and Austin, L. Depression simulating organic brain disease. Am. J. Psychiat. 136:521–523, 1979.
24. Corkin, S. Acetylcholine, aging and Alzheimer's disease. Implications for treatment. Trends in NeuroSci. 4:287–290, 1981.

25. Corsellis, J. A. N. *Mental Illness in the Aging Brain*. London: Oxford Univ. Press, 1962.
26. Cross, A. J., Crow, T. J., Perry, E. K., Perry, R. H., Blessed, G., and Tomlinson, B. E. Reduced dopamine-beta-hydroxylase activity in Alzheimer's disease. Brit. Med. J. 282:93–94, 1981.
27. Cummings, J., Benson, D. F., and LoVerme, S., Jr. Reversible dementia. Illustrative cases, definition, and review. JAMA 243:2434–2439, 1980.
28. Cutler, N. R., Heston, L. L., Davies, P., Haxby, J. V., and Schapiro, M. B. Alzheimer's disease and Down's syndrome: new insights. Ann. Intern. Med. 103:566–578, 1985.
29. Dahl, D., Selkoe, D. J., Pero, R. T., and Bignami, A. Immunostaining of neurofibrillary tangles in Alzheimer's senile dementia with a neurofilament antiserum. J. Neurosci. 2:113–119, 1982.
30. Davies, P., and Maloney, A. J. F. Selective loss of central cholinergic neurons in Alzheimer's disease. Lancet 2:1403, 1976.
31. Davis, K. L., Mohs, R. C., and Tinklenberg, J. R. Enhancement of memory by physostigmine. N. Engl. J. Med. 301:946, 1979. (Letter to the editor.)
32. de Alarcón, J. Social causes and social consequences of mental illness in old age. In *Recent Developments in Psychogeriatrics*, Kay, D. W. K., and Walk, A. (eds.) Ashford, Kent: Headley Brothers, 1971.
33. Eagles, J. M., Craig, A., Rawlinson, F., Restall, D. B., Beattie, J. A. G., and Besson, J. A. O. The psychological well-being of supporters of the demented elderly. Brit. J. Psychiat. 150:293–298, 1987.
34. Eagles, J. M., Beattie, J. A. G., Blackwood, G. W., Restall, D. B., and Ashcroft, G. W. The mental health of elderly couples. I. The effects of a cognitively impaired spouse. Brit. J. Psychiat. 150:299–303, 1987.
35. Eagles, J. M., Walker, L. G., Blackwood, G. W., Beattie, J. A. G., and Restall, D. B. The mental health of elderly couples. II. Concordance for psychiatric morbidity in spouses. Brit. J. Psychiat. 150:303–308, 1987.
36. Easson, W. Myxedema with psychosis. Arch. Gen. Psychiat. 14:277–283, 1966.
37. Editorial. Cholinergic involvement in senile dementia. Lancet 1:408–409, 1977.
38. Enna, S. J., Bird, E. D. Bennett, J. P. Bylund, D. B., Yamamura, H. I., Iversen, L. L., and Snyder, S. H. Huntington's chorea: changes in neurotransmitter receptors in the brain. N. Engl. J. Med. 294:1305–1309, 1976.
39. Folstein, S. E., Franz, M. L., Jensen, B. A., Chase, G. A., and Folstein, M. F. Conduct disorder and affective disorder among the offspring of patients with Huntington's disease. Psychol. Med. 13:45–52, 1983.
40. Folstein, S. E., Abbott, M. H., Chase, G. A., Jensen, B. A., and Folstein, M. F. The association of affective disorder with Huntington's disease in a case series and in families. Psychol. Med. 13:537–542, 1983.
41. Frackowiak, R. S. J., Pozzilli, C., Legg, N. J., du Boulay, G. H., Marshall, J., Lenzi, G. L., and Jones, T. Regional cerebral oxygen supply

and utilization in dementia. A clinical and physiological study with oxygen-15 and positron tomography. Brain 104:753–778, 1981.

42. Fuller, J., Ward, E., Evans, A., Massam, K., and Gardner, A. Dementia: supportive groups for relatives. Brit. Med. J. 1:1684–1685, 1979.

43. Gjestland, T. The Oslo study of untreated syphilis. Acta Derm. Venereol. Suppl. 34, 1955.

44. Goldgaber, D., Lerman, M. I., McBride, O. W., Saffiotti, U., and Gajdusek, D. C. Characterization and chromosomal localization of a cDNA encoding brain amyloid of Alzheimer's disease. Science 235:877–880, 1987.

45. Gronwall, D., and Wrightson, P. Cumulative effect of concussion. Lancet 2:995–997, 1975.

46. Grundke-Iqbal, I., Johnson, A. B., Wisniewski, H. M., Terry, R. D., and Iqbal, K. Evidence that Alzheimer neurofibrillary tangles originate from neurotubules. Lancet 1:578–580, 1979.

47. Gusella, J. F., Wexler, N. S., Conneally, P. M., Naylor, S. L., Anderson, M. A., Tanzi, R. E., Watkins, P. C., Ottina, K., Wallace, M. R., Sakaguchi, A. Y., Young, A. B., Shoulson, I., Bonilla, E., and Martin, J. B. A polymorphic DNA marker genetically linked to Huntington's disease. Nature 306:234–238, 1983.

48. Guze, S. B., and Cantwell, D. P. The prognosis in "organic brain" syndromes. Am. J. Psychiat. 120:878–881, 1964.

49. Guze, S. B., and Daengsurisri, S. Organic brain syndromes. Prognostic significance in general medical patients. Arch. Gen. Psychiat. 17:365–366, 1967.

50. Harper, P. S., Tyler, A., Smith, S., Jones, P., Newcombe, R. G., and McBroom, V. Decline in the predicted incidence of Huntington's chorea associated with systematic genetic counselling and family support. Lancet 2:411–413, 1981.

51. Hendrie, H. C., ed. Brain Disorders: Clinical Diagnosis and Management. Psychiat. Clin. North America, Vol. 1, April 1978.

52. Heston, L. L., and Mastri, A. R. The genetics of Alzheimer's disease. Arch. Gen. Psychiat. 34:976–981, 1977.

53. Heston, L. L. Alzheimer's disease, trisomy 21, and myeloproliferative disorders: associations suggesting a genetic diathesis. Science 196:322–323, 1977.

54. Heston, L. L., Mastri, A. R., Anderson, V. E., and White, J. Dementia of the Alzheimer type. Arch. Gen. Psychiat. 38:1085–1090, 1981.

55. Heston, L. L. Down's syndrome and Alzheimer's dementia: defining an association. Psychiat. Dev. 4:287–294, 1984.

56. Ho, D. D., Rota, T. R., Schooley, R. T., Kaplan, J. C., Allan, J. D., Groopman, J. E., Resnick, L., Felsenstein, D., Andrews, C. A., and Hirsch, M. S. Isolation of HTLV-III from cerebrospinal fluid and neural tissues of patients with neurologic syndromes related to the acquired immunodeficiency syndrome. N. Engl. J. Med. 313:1493–1497, 1985.

57. Hudson, A. J. Amyotrophic lateral sclerosis and its association with dementia, parkinsonism and other neurological disorders: a review. Brain 104:217–247, 1981.
58. Hunter, R., and Macalpine, I. Three Hundred Years of Psychiatry. 1535–1860. London: Oxford Univ. Press, 1963.
59. Jacobs, L., Conti, D., Kinkel, W. R., and Manning, E. J. "Normal-pressure" hydrocephalus: relationship of clinical and radiographic findings to improvement following shunt surgery. JAMA 235:510–512, 1976.
60. Kahn, R. L., Zarit, S. H., Hilbert, N. M., and Niederehe, G. Memory complaint and impairment in the aged. Arch. Gen. Psychiat. 32:1569–1573, 1975.
61. Kim, R. C., Collins, G. H., Parisi, J. E., Wright, A. W., and Chu, Y. B. Familial dementia of adult onset with pathological findings of a "non-specific" nature. Brain 104:61–78, 1981.
62. Larsson, T., Sjögren, T., and Jacobson, G. Senile dementia. Acta Psychiat., Suppl. 167, 1963.
63. Lazarus, L. W., Newton, N., Cohler, B., Lesser, J., and Schweon, C. Frequency and presentation of depressive symptoms in patients with primary degenerative dementia. Am. J. Psychiat. 144:41–45, 1987.
64. Leading Article. Communicating hydrocephalus. Lancet 2:1011–1012, 1977.
65. Lishman, W. A. Cerebral disorder in alcoholism. Syndromes of impairment. Brain 104:1–20, 1981.
66. Liston, E. H. Clinical findings in presenile dementia. A report of 50 cases. J. Nerv. Ment. Dis. 167:337–342, 1979.
67. Liston, E. H. The clinical phenomenology of presenile dementia. A critical review of the literature. J. Nerv. Ment. Dis. 167:329–336, 1979.
68. Locke, B. Z., Kramer, M., and Pasamanick, B. Mental disorders of the senium at mid-century: first admissions to Ohio State public mental hospitals. Am. J. Pub. Health. 50:998–1012, 1960.
69. Lyle, O. E., and Gottesman, I. I. Subtle cognitive deficits as 15- to 20-year precursors of Huntington's disease. In Advances in Neurology, Chase, T. N. et al. (eds.), 23:227–238, New York: Raven Press, 1979.
70. Martin, J. B., and Gusella, J. F. Huntington's disease. Pathogenesis and management. N. Engl. J. Med. 315:1267–1276, 1986.
71. McAllister, T. W., and Price, T. R. P. Severe depressive pseudodementia with and without dementia. Am. J. Psychiat. 139:626–629, 1982.
72. McDermott, J. R., Smith, A. I., Ward, M. K., Parkinson, I. S., and Kerr, D. N. S. Brain-aluminium concentration in dialysis encephalopathy. Lancet 1:901–904, 1978.
73. Mohs, R. C., Breitner, J. C. S., Silverman, J. M., and Davis, K. L. Alzheimer's disease. Morbid risk among first-degree relatives approximates 50% by 90 years of age. Arch. Gen. Psychiat. 44:405–408, 1987.
74. Morgagni, G. B. The Seats and Causes of Diseases Investigated by Anatomy. (Trans. B. Alexander). Venice, 1761. London: Millar et al., 1769.
75. Morris, J. C., Rubin, E. H., Morris, E. J., and Mandel, S. A. Senile

dementia of the Alzheimer's type: an important risk factor for serious falls. J. Gerontol. 42:412–417, 1987.

76. Navia, B. A., Jordan, B. D., and Price, R. W. The AIDS dementia complex: I. Clinical features. Ann. Neurol. 19:517–524, 1986.

77. Potamianos, G., and Kellett, J. M. Anti-cholinergic drugs and memory: the effects of benzhexol on memory in a group of geriatric patients. Brit. J. Psychiat. 140:470–472, 1982.

78. Pratt, R. T. C. The Genetics of Neurological Disorders. New York: Oxford Univ. Press, 1967.

79. Preskorn, S. H., and Simpson, S. Tricyclic-antidepressant-induced delirium and plasma drug concentration. Am. J. Psychiat. 139:822–823, 1982.

80. Price, D. L., Whitehouse, P. J., Struble, R. G., Clark, A. W., Coyle, J. T., DeLong, M. R., and Hedreen, J. C. Basal forebrain cholinergic systems in Alzheimer's disease and related dementias. Neurosci. Comm. 1:84–92, 1982.

81. Rabins, P. V., and Folstein, M. F. Delirium and dementia: diagnostic criteria and fatality rates. Brit. J. Psychiat. 140:149–153, 1982.

82. Report from the Registration Committee of the European Dialysis and Transplant Association. Dialysis dementia in Europe. Lancet 2:190–192, 1980.

83. Rice, E., and Gendelman, S. Psychiatric aspects of normal pressure hydrocephalus. JAMA 223:409–412, 1973.

84. Robins, E., Murphy, G. E., Wilkinson, R. H., Jr., Gassner, S., and Kayes, J. Some clinical considerations in the prevention of suicide based on a study of 134 successful suicides. Am. J. Pub. Health 49:888–899, 1959.

85. Ron, M. A. Brain damage in chronic alcoholism: a neuropathological, neuroradiological and psychological review. Psychol. Med. 7:103–112, 1977.

86. Ron, M.A., Toone, B. K., Garralda, M. E., and Lishman, W. A. Diagnostic accuracy in presenile dementia. Brit. J. Psychiat. 134:161–168, 1979.

87. Rossor, M. N., Garrett, N. J., Johnson, A. L., Mountjoy, C. Q., Roth, M., and Iversen, L. L. A post-mortem study of the cholinergic and GABA systems in senile dementia. Brain 105:313–330, 1982.

88. Roth, M. Classification and etiology in mental disorders of old age. In Recent Developments in Psychogeriatrics, Kay, D. W. K. and Walk, A. (eds.). Ashford, Kent: Headley Brothers, 1971.

89. Salmon, W. Iatrica: Sen Praxis Medendi. The Practice of Curing Disease, 3d ed. London: Rolls, 1964.

90. Schafer, D. F., and Jones, E. A. Hepatic encephalopathy and the γ-amino-butyric-acid neurotransmitter system. Lancet 1:18–20, 1982.

91. Schentag, J. J., Cerra, F. B., Calleri, G., DeGlopper, E. Rose, J. Q., and Bernhard, H. Pharmacokinetic and clinical studies in patients with cimetidine-associated mental confusion. Lancet 1:177–181, 1979.

92. Schneck, M. K., Reisberg, B., and Ferris, S. H.   An overview of current concepts of Alzheimer's disease. Am. J. Psychiat. 139:165–173, 1982.
93. Shaw, G. M., Harper, M. E., Hahn, B. H., Epstein, L. G., Gajdusek, D. C., Price, R. W., Navia, B. A., Petito, C. K., O'Hara, C. J., Groopman, J. E., Cho, E.-S., Oleske, J. M., Wong-Staal, F., and Gallo, R. C. HTLV-III infection in brains of children and adults with AIDS encephalopathy. Science 227:177–182, 1985.
94. Smith, C. K., Barish, J., Correa, J., and Williams, R. H.   Psychiatric disturbance in endocrinologic disease. Psychosom. Med. 34:69–86, 1972.
95. Sternberg, D. E., and Jarvik, M. E.   Memory functions in depression. Arch. Gen. Psychiat. 33:219–224, 1976.
96. Stoler, M. H., Eskin, T. A., Benn, S., Angerer, R. C., and Angerer, L. M.   Human T-cell lymphotropic virus type III infection of the central nervous system. JAMA 256:2360–2364, 1986.
97. Strachan, R., and Henderson, J.   Psychiatric syndromes due to avitaminosis B_{12} with normal blood and marrow. Q. J. Med. 34:303–317, 1965.
98. Tanzi, R. E., Gusella, J. F., Watkins, P. C., Bruns, G. A. P., St. George-Hyslop, P., Van Keuren, M. L., Patterson, D., Pagan, S., Kurnit, D. M., and Neve, R. L.   Amyloid β protein gene: cDNA, mRNA distribution, and genetic linkage near the Alzheimer locus. Science 235:880–884, 1987.
99. Thomas, P. K.   Brain atrophy and alcoholism. Brit. Med. J. 292:787, 1986.
100. Tomlinson, B. E.   Plaques, tangles and Alzheimer's disease. Psychol. Med. 12:449–459, 1982. (Editorial.)
101. Trimble, M. R., and Reynolds, E. H.   Anticonvulsant drugs and mental symptoms: a review. Psychol. Med. 6:169–178, 1976.
102. Tune, L. E., Damlouji, N. F., Holland, A., Gardner, T. J. Folstein, M. F., and Coyle, J. T.   Association of postoperative delirium with raised serum levels of anticholinergic drugs. Lancet 2:651–653, 1981.
103. Victor, M., Adams, R. D., and Collins, G. H.   The Wernicke-Korsakoff Syndrome. Philadelphia: F. A. Davis, 1971.
104. Victoratos, G. C., Lenman, J. A. R., and Herzberg, L.   Neurological investigation of dementia. Brit. J. Psychiat. 130:131–133, 1977.
105. Weinberger, D. R.   Brain disease and psychiatric illness: when should a psychiatrist order a CAT scan? Am. J. Psychiat. 141:1521–1527, 1984.
106. Well, C. E.   Chronic brain disease: an overview. Am. J. Psychiat. 135:1–12, 1978.
107. Wells, C. E.   Pseudodementia. Am. J. Psychiat. 136:895–900, 1979.
108. Whalley, L. J., Carothers, A. D., Collyer, S., DeMay, R., and Frackiewicz, A.   A study of familial factors in Alzheimer's disease. Brit. J. Psychiat. 140:249–256, 1982.
109. Whitehouse, P. J., Price, D. L., Struble, R. G., Clark, A. W., Coyle, J. T., and DeLong, M. R.   Alzheimer's disease and senile dementia: loss of neurons in the basal forebrain. Science 215:1237–1239, 1982.

11. Anorexia Nervosa

Definition

Anorexia nervosa is characterized by peculiar attitudes toward eating and weight that lead to obsessive refusal to eat, profound weight loss, and, when the disorder occurs among girls, persistent amenorrhea. Bulimia refers to the gorging of food followed by induced vomiting. It is seen in many patients with anorexia nervosa and also may occur alone or as the predominant clinical feature.

Historical Background

In 1689 Richard Morton published a monograph entitled *Phthisiologia or a Treatise of Consumptions*. One of its early chapters, "Nervous Phthisis," contains case histories of the illness we recognize today as anorexia nervosa. An example follows:

Mr Duke's Daughter is S. Mary Axe, in the Year 1684 and the Eighteenth Year of her Age, in the Month of July fell into a total Suppression of her Monthly Courses from a multitude of Cares and Passions of her Mind, but without any Symptoms of the Green-Sickness following upon it. From which time her Appetite began to abate, and her Digestion to be bad; her Flesh also began to be flaccid and loose, and her Looks pale, with other symptoms usual in Universal Consumption of the Habit of the Body and by the extreme and memorable cold weather which happened the Winter following, this Consumption did seem to be not a little improved; for that she was wont by her studying at Night, and continual poring upon Books, to expose herself both Day and Night to the Injuries of the Air, which was at that time extremely cold, not without some manifest Prejudice to the System of her Nerves. The Spring following, by the Prescription of some Empirik, she took a Vomit, and after that I know not what Steel Medicine, but without any Advantage. So from that time loathing all sorts of Medicaments, she wholly neglected the care of herself for two full Years, till at last being brought to the last degree of Marasmus, or Consumption, and thereupon subject to frequent Fainting-Fitts, she apply'd herself to me for Advice.

I do not remember that I did ever in all my practice see one, that was conversant with the Living so much wasted with the greatest degree of Consumption, (like a Skeleton only clad with Skin) yet there was no Fever, but on the contrary a Coldness of the whole Body; no cough, or Difficulty of Breathing nor an appearance of any other distemper of the Lungs, or of any other Entrail; No Looseness, or any other sign of a Colliquation, or Preternatural Expence of the Nutritious Juices. Only her Appetite was diminished, and her Digestion uneasy, with Fainting-Fitts, which did frequently return upon her. Which symptoms I did endeavor to relieve by the outward Application of Aromatick Bags made to the Region of the Stomach, and by Stomach-Plaisters, as also by the internal use of bitter Medicines, Chalybeates, and Juleps made of Cephalick and Antihysterick Waters, sufficiently impregnated with Spirit of Salt Armoniack, and Tincture of Castor, and other things of that Nature. Upon the use of which she seemed to be much better; but being quickly tired with Medicines, she beg'd that the whole Affair might be committed again to Nature, whereupon, consuming every day more and more, she was after three Months taken with a Fainting-Fitt, and died.

Another vivid and accurate description was presented in 1908 by Dejerine and Gauckler (18):

It sometimes happens that a physician has patients—they are more apt to be women—whose appearance is truly shocking. Their eyes are brilliant. Their cheeks are hollow, and their cheek bones seem to protrude through the skin. Their withered breasts hang from the walls of their chest. Every rib stands out. Their shoulder blades appear to be loosened from their frame. Every vertebra shows through the skin. The abdominal wall sinks in below the floating ribs and forms a hollow like a basin. The thighs and the calves of their legs are reduced to a skeleton. One would say it was the picture of an immured nun, such as the old masters have portrayed. These women appear to be fifty or sixty years old. Sometimes they seem to be sustained by some unknown miracle of energy; their voices are strong and their steps firm. On the other hand, they often seem almost at the point of death, and ready to draw their last breath.

Are they tuberculous or cancerous patients, or muscular atrophies in the last stages, these women whom misery and hunger have reduced to this frightful gauntness? Nothing of the kind. Their lungs are healthy, there is no sign of any organic affection. Although they look so old they are young women, girls, sometimes children. They may belong to good families, and be surrounded by every care. These patients are what are known as mental anorexics, who, without having any physical lesions, but by the association of various troubles, all having a psychic origin, have lost a quarter, a third, and sometimes a half of their weight. The affection which has driven them to this point may have lasted months, sometimes years. Let it go on too long and death will occur, either from inanition or from secondary tuberculosis. However, it is a case of nothing but a purely psychic affection of which the mechanisms are of many kinds.

Gull and Lasègue described the syndrome independently in 1873 (29, 55). The English term for the illness has been "anorexia nervosa," the French "*l'anorexie hystérique*," and the German "*Pubertaetsmagersucht*." Of these, the last is probably most precise; its literal meaning is "adolescent pursuit of thinness." It is this relentless pursuit of thinness that especially distinguishes anorexia nervosa from other illnesses associated with loss of weight.

Older historical accounts of patients wasting to the point of death do not usually provide enough information to permit distinction between cases of anorexia nervosa, tuberculosis, or panhypopituitarism. Until relatively recently it was not possible to make the distinction with confidence. Some early photographs of alleged pituitary disease, illustrating markedly wasted patients, probably represent cases of anorexia nervosa (see Differential Diagnosis).

The cause of the illness is not known, though hypotheses involving endocrine, hypothalamic, and psychosocial factors abound (16, 32, 39, 61). One frequently invoked hypothesis is that the syndrome represents a rejection of adult sexuality with associated fears of oral impregnation. A wide spectrum of sexual knowledge, attitudes, and behavior was found in one study of patients with anorexia nervosa, though many believed that their illnesses had been precipitated by some sexual experience (4).

Epidemiology

Anorexia nervosa is not a common illness. On the basis of data from three community psychiatric case registers, in Scotland, England, and the United States, the average annual incidence appears to be low, in the range of one case per one hundred thousand population (15, 51). The Maudsley Hospital in London admitted thirty-eight patients over a twenty year period (50). Most large teaching hospitals admit several cases each year. Recent evidence suggests, however, that the incidence of anorexia nervosa may be increasing (47), particularly the type associated with gorging of food followed by self-induced vomiting (22, 35, 67, 75).

Anorexia nervosa occurs much more frequently in girls than in boys. Probably 90 percent of cases are girls (38, 60). Obviously, if amenorrhea is required for the diagnosis, the disorder will be limited to females. Some authors believe anorexia nervosa is mainly an illness of middle-class and upper-class girls, but there are no controlled data to support this hypothesis. Anorexia nervosa appears to be rare in blacks, though there is a suggestion that such cases may be increasing (69). It is apparently overrepresented among professional dance and modeling students, usually developing after the students have begun their studies (24, 27).

Clinical Picture

The syndrome typically begins with concern about mild obesity, followed by negative attitudes toward eating. A disgust for food that is far stonger than hunger develops. As weight loss progresses, patients may lose their hunger. However thin, they still consider themselves fat and continue to lose weight. These patients may lose so much weight that they resemble survivors of concentration camps, but with several differences. Patients with anorexia nervosa are usually alert and cheerful. In addition, they may be hyperactive. They often engage in strenuous exercise, sometimes in an open effort to keep from gaining weight.

From the onset of the illness, odd behavior relating to food may be seen. Patients may gorge themselves (bulimia), then induce vomiting voluntarily. This feature has attracted considerable attention in recent years, and, in some patients, it may be the dominant clinical problem (12, 22, 26, 35, 62). Although starving themselves, patients may hoard food secretly or throw it away. To reduce weight, they may abuse laxatives, enemas, and diuretics.

Amenorrhea may begin before, with, or after the disturbance of appetite (23). It is present in most cases. A few patients develop anorexia nervosa after they have had the experience of motherhood, indicating that an undeveloped reproductive system is not an essential feature of the illness. Other physical findings are bradycardia and lanugo (soft, downy body hair).

Physiological differences between patients with anorexia nervosa and others have been noted (1, 9, 42, 58, 63). Most investigators believe that starvation produces these abnormalities, which include an abnormal glucose tolerance curve; increased serum cholesterol; increased serum carotene levels; diminished urinary 17-ketosteroids, estrogens, and gonadotropins; and low basal metabolic rate (16, 44). A striking "immaturity" in the pattern of luteinizing hormone secretion has been described in which the patients' hormone functioning resembles that of prepubertal girls but returns to normal after remission of the anorexia nervosa (8, 48). Male anorexia nervosa patients are reported to have low testosterone levels (6).

Many reports suggest an association between anorexia nervosa and Turner's syndrome or other congenital gonadal dysgenesis (17, 19, 33, 40, 53, 54), but the data are still not conclusive (73).

DSM-III-R criteria for the diagnosis of anorexia nervosa and its bulimic variant are presented in Tables 11.1 and 11.2.

Natural History

There has been disagreement about whether anorexia nervosa is a specific illness or a manifestation of other illnesses (50). However, if patients who have a preexisting psychiatric illness other

Table 11.1 Diagnostic criteria for Anorexia Nervosa (DSM-III-R)

A. Refusal to maintain body weight over a minimal normal weight for age and height, e.g., weight loss leading to maintenance of body weight 15% below that expected; or failure to make expected weight gain during period of growth, leading to body weight 15% below that expected.
B. Intense fear of gaining weight or becoming fat, even though underweight.
C. Disturbance in the way in which one's body weight, size, or shape is experienced, e.g., the person claims to "feel fat" even when emaciated, believes that one area of the body is "too fat" even when obviously underweight.
D. In females, absence of at least three consecutive menstrual cycles when otherwise expected to occur (primary or secondary amenorrhea). (A woman is considered to have amenorrhea if her periods occur only following hormone, e.g., estrogen, administration.)

---

than anorexia nervosa are separated from those whose history begins with the symptoms of anorexia nervosa, a pure group can be isolated (52).

The age of onset ranges from prepuberty to young adulthood, with the mean in the midteens (74). Premorbidly, patients with anorexia nervosa tend to be shy and introverted (16). A history of early feeding difficulties is common, as is a history of obsessional traits (16). The onset may be abrupt. There is one re-

Table 11.2 Diagnostic criteria for Bulimia Nervosa (DSM-III-R)

A. Recurrent episodes of binge eating (rapid consumption of a large amount of food in a discrete period of time).
B. A feeling of lack of control over eating behavior during the eating binges.
C. The person regularly engages in either self-induced vomiting, use of laxatives or diuretics, strict dieting or fasting, or vigorous exercise in order to prevent weight gain.
D. A minimum average of two binge eating episodes a week for at least three months.
E. Persistent overconcern with body shape and weight.

---

corded case of an interval of six months from onset to the time of death (50). The illness may involve one lengthy episode lasting many months or years or it may be marked by remissions and exacerbations.

In one follow-up study, full recovery occurred in half the patients. Other patients had persistent difficulties, even when the problem of initial massive weight loss was resolved. These included menstrual dysfunction, sexual maladjustment, large weight fluctuations, and disturbed appetite (3, 13, 31, 43, 50, 66, 70, 71). When full recovery occurs, it is usually within three years of the onset (16). Depressive symptoms are common (11, 21); occasionally patients describe previous periods of anorexia nervosa so as to make them sound like episodes of depression. Dexamethasone suppression of cortisol secretion is impaired in some patients with anorexia nervosa, particularly those with considerable weight loss (28). This raises the possibility that impaired dexamethasone suppression in some patients with depression may be due to the associated weight loss or that some cases of anorexia nervosa may be manifestations of an underlying affective illness.

Sometimes the illness is associated with possible precipitating events such as the death of a relative or a broken engagement (4, 50). On the other hand, many patients can give no reason for the onset of their dieting (16).

A better prognosis appears to be related to earlier age at onset, though not all have reported this finding (37), and to greater educational achievement. A poorer prognosis appears to be related to later age at onset, longer duration of illness, disturbed relationship with parents, vomiting, laxative abuse, and the severity of obsessionality and depressive symptoms (5, 34, 56, 68).

The extent of behavioral disturbance is not related to outcome. Thus, a patient who is adamant in her refusal to eat is no more likely to have a poor prognosis than one who cooperates with treatment (16).

Though the immediate response to treatment may be good, premature death occurs in up to 10 or 15 percent of cases hospitalized psychiatrically; sometimes patients die a few years after the onset of their illness. In one series, there was a 5 percent mortality within four years (56). Thus, anorexia nervosa should not be regarded lightly. When death occurs, the causes are commonly starvation and its complications, but it is often not possible to determine the cause of death (59).

Recent reports suggest that the course and outcome of anorexia nervosa in males does not differ significantly from that in females (10, 72).

Complications

The most obvious complication of anorexia nervosa is death from starvation.

Patients with anorexia nervosa apparently develop psychoses no more often than do persons in the general population (16). When psychotic features do develop, they usually appear to be part of associated major affective disorders (45).

Family Studies

Some clinicians say that anorexia patients have dominating mothers and passive fathers (52); others disagree, finding neither a single personality pattern among parents nor a single pattern of interaction among family members (16, 25, 30). Close relatives of patients with anorexia nervosa do not have an increased prevalence of schizophrenia (73), but the findings concerning familial affective disorders are contradictory, as are the data concerning the association between anorexia nervosa and affective disorder in index patients (2, 46, 49, 57). There is evidence of increased parental "neurosis" (16, 50, 52), but anorexia nervosa is rare, though in one study (68) an increased prevalence among sisters was found. In another study (36), an increased familial prevalence of alcoholism was noted.

Of four pairs of monozygotic twins that have been reported, three were concordant for anorexia nervosa and one discordant (33, 64). A more recent series of thirty-four twin pairs and one set of triplets in which the proband had anorexia nervosa showed a striking increase in concordance among female monozygotic twins compared to dizygotic twins, but no concordance in the small number of male twins (41).

Differential Diagnosis

Anorexia nervosa is characterized by unceasing pursuit of thinness and should not be confused with weight loss occurring in the course of illness such as depression or schizophrenia.

Starvation, when it has causes other than anorexia nervosa, is usually accompanied by apathy and inactivity rather than by the alertness and hyperactivity characteristic of anorexia nervosa.

Patients with anorexia nervosa or panhypopituitarism may both have low levels of urinary gonadotrophins, but they are not as consistently low in the former as in the latter (50). Hypopituitarism is seldom associated with the severe cachexia of anorexia nervosa or the strikingly high level of physical activity.

Clinical Management

Treatment of patients with anorexia nervosa is far from satisfactory, and there is no agreement about the best form. Some patients require constant observation on a locked psychiatric floor and tube feeding. Others have been successfully treated with phenothiazines (14). Quite recently, techniques of behavior modification have been used with reported success (7, 20, 65). It is too early in the course of investigations of the treatment of anorexia nervosa by methods based on learning theory to evaluate their usefulness.

References*

1. Ainley, C. C., Cason, J., Carlsson, L., Thompson, R. P. H., Slavin, B. M., and Norton, K. R. W. Zinc state in anorexia nervosa. Brit. Med. J. 293:992–993, 1986.

* See also the extensive review of the field presented at an International Conference on Anorexia Nervosa and Related Disorders. J. Psychiat. Res. 19:79–521, 1985.

2. Altshuler, K. Z., and Weiner, M. F.   Anorexia nervosa and depression: a dissenting view. Am. J. Psychiat. 142:328–332, 1985.

3. Bassø, H. H., and Eskeland, I.   A prospective study of 133 patients with anorexia nervosa. Acta Psychiat. Scand. 65:127–133, 1982.

4. Beumont, P. J. V., Abraham, S. F., and Simson, K. G.   The psychosexual histories of adolescent girls and young women with anorexia nervosa. Psychol. Med. 11:131–140, 1981.

5. Beumont, P. J. V., George, G. C. W., and Smart, D. E.   "Dieters" and "vomiters and purgers" in anorexia nervosa. Psychol. Med. 6:617–622, 1976.

6. Beumont, P. J. V., Veardwood, C. J., and Russell, G. F. M.   The occurrence of the syndrome of anorexia nervosa in male subjects. Psychol. Med. 2:216–231, 1972.

7. Blinder, B. J., Freeman, D. M. A., and Stunkard, A. J.   Behavior therapy of anorexia nervosa: effectiveness of activity as a reinforcer of weight gain. Am. J. Psychiat. 126:1093–1099, 1970.

8. Boyar, R. M., Katz, J., Finkelstein, J. W., Kapen, S., Wiener, H., Weitzman, E. D., and Hellman. L.   Anorexia nervosa: immaturity of the 24-hour luteinizing hormone secretory pattern. N. Engl. J. Med. 291:861–865, 1974.

9. Brotman, A. W., Rigotti, N., and Herzog, D. B.   Medical complications of eating disorders: outpatient evaluation and management. Compr. Psychiat. 26:258–272, 1985.

10. Burns, T., and Crisp. A. H.   Outcome of anorexia nervosa in males. Brit. J. Psychiat. 145:319–325, 1984.

11. Cantwell, D. P., Sturzenberger, S., Burroughs, J., Salkin, B., and Green, J. K.   Anorexia nervosa: an affective disorder? Arch. Gen. Psychiat. 34:1087–1093, 1977.

12. Casper, R. C., Eckert, E. D., Halmi, K. A., Goldberg, S. C., and Davis, J. M.   Bulimia. Its incidence and clinical importance in patients with anorexia nervosa. Arch. Gen. Psychiat. 37:1030–1035, 1980.

13. Casper, R. C., Halmi, K. A., Goldberg, S. C., Eckert, E. D., and Davis, J. M.   Disturbances in body image estimation as related to other characteristics and outcome in anorexia nervosa. Brit. J. Psychiat. 134:60–66, 1979.

14. Crisp, A. H.   A treatment regimen for anorexia nervosa. Brit. J. Psychiat. 112:505, 1966.

15. Crisp, A. H., Palmer, R. L., and Kalucy, R. S.   How common is anorexia nervosa? A prevalence study. Brit. J. Psychiat. 128:549–554, 1976.

16. Dally, P.   Anorexia Nervosa. New York: Grune & Stratton, 1969.

17. Darby, P. L., Garfinkel, P. E., Vale, J. M., Kirwan, P. J., and Brown, G. M.   Anorexia nervosa and "Turner syndrome": cause or coincidence? Psychol. Med. 11:141–145, 1981.

18. Dejerine, J., and Gauckler, E.   Le Réeducation des faux gastropathies. Presse Méd. 16:225, 1908.

19. Dougherty, G. G., Jr., Rockwell, W. J. K., Sutton, G., and Ellinwood,

E. H., Jr. Anorexia nervosa in treated gonadal dysgenesis: case report and review. J. Clin. Psychiat. 44:219–221, 1983.

20. Eckert, E. D., Goldberg, S. C., Halmi K. A., Casper, R. C., and Davis, J. M. Behaviour therapy in anorexia nervosa. Brit. J. Psychiat. 134:55–59, 1979.

21. Eckert, E. D., Goldberg, S. C., Halmi, K. A., Casper, R. C., and Davis, J. M. Depression in anorexia nervosa. Psychol. Med. 12:115–122, 1982.

22. Fairburn, C. G., and Cooper, P. J. Self-induced vomiting and bulimia nervosa: an undetected problem. Brit. Med. J. 284:1153–1155, 1982.

23. Falk, J. R., and Halmi, K. A. Amenorrhea in anorexia nervosa: examination of the critical body weight hypothesis. Biol. Psychiat. 17:799–806, 1982.

24. Frisch, R. E., Wyshak, G., and Vincent, L. Delayed menarche and amenorrhea in ballet dancers. N. Engl. J. Med. 303:17–19, 1980.

25. Garfinkel, P. E., Garner, D. M., Rose, J., Darby, P. L., Brandes, J. S., O'Hanlon, J., and Walsh, N. A comparison of characteristics in the families of patients with anorexia nervosa and normal controls. Psychol. Med. 13:821–828, 1983.

26. Garfinkel, P. E., Moldofsky, H., and Garner, D. M. The heterogeneity of anorexia nervosa. Arch. Gen. Psychiat. 37:1036–1040, 1980.

27. Garner, D. M., and Garfinkel, P. E. Socio-cultural factors in the development of anorexia nervosa. Psychol. Med. 10:647–656, 1980.

28. Gerner, R. H., and Gwirtsman, H. E. Abnormalities of dexamethasone suppression test and urinary MHPG in anorexia nervosa. Am. J. Psychiat. 138:650–653, 1981.

29. Gull, W. W. Anorexia nervosa (apepsia hysterica). Brit. Med. J. 2:527, 1873.

30. Hall, A., Leibrich, J., Walkey, F. H., and Welch, G. Investigation of "weight pathology" of 58 mothers of anorexia nervosa patients and 204 mothers of schoolgirls. Psychol. Med. 16:71–76, 1986.

31. Hall, A., Slim, E., Hawker, F., and Salmond, C. Anorexia nervosa: long-term outcome in 50 female patients. Brit. J. Psychiat. 145:407–413, 1984.

32. Halmi, K. A. The state of research in anorexia nervosa and bulimia. Psychiat. Dev. 3:247–262, 1983.

33. Halmi, K., and Brodland, G. Monozygotic twins concordant and discordant for anorexia nervosa. Psychol. Med. 3:521–524, 1973.

34. Halmi, K., Brodland, G., and Loney, J. Prognosis in anorexia nervosa. Ann. Intern. Med. 78:907–909, 1973.

35. Halmi, K. A., Falk, J. R., and Schwartz, E. Binge-eating and vomiting: a survey of a college population. Psychol. Med. 11:697–706, 1981.

36. Halmi, K. A., and Loney, J. Familial alcoholism in anorexia nervosa. Brit. J. Psychiat. 123:53–54, 1973.

37. Hawley, R. M. The outcome of anorexia nervosa in younger subjects. Brit. J. Psychiat. 146:657–660, 1985.

38. Hay, G. G., and Leonard, J. C. Anorexia nervosa in males. Lancet 2:574–576, 1979.

39. Herzog, D. B., and Copeland, P. M. Eating disorders. N. Engl. J. Med. 313:295–303, 1985.
40. Hindler, C. G., and Norris, D. L. A case of anorexia nervosa with Klinefelter's syndrome. Brit. J. Psychiat. 149:659–660, 1986.
41. Holland, A. J., Hall, A., Murray, R., Russell, G. F. M., and Crisp, A. H. Anorexia nervosa: a study of 34 twin pairs and one set of triplets. Brit. J. Psychiat. 145:414–419, 1984.
42. Holt, S., Ford, M. J., Grant, S., and Heading, R. C. Abnormal gastric emptying in primary anorexia nervosa. Brit. J. Psychiat. 139:550–552, 1981.
43. Hsu, L. K. G. Outcome of anorexia nervosa. Arch. Gen. Psychiat. 37: 1041–1046, 1980.
44. Hudson, J. I., and Hudson, M. S. Endocrine dysfunction in anorexia nervosa and bulimia: comparison with abnormalities in other psychiatric disorders and disturbances due to metabolic factors. Psychiat. Dev. 4:237–272, 1984.
45. Hudson, J. I., Pope, H. G., Jr., and Jonas, J. M. Psychosis in anorexia nervosa and bulimia. Brit. J. Psychiat. 145:420–423, 1984.
46. Hsu, L. K. G., Holder, D., Hindmarsh, D., and Phelps, C. Bipolar illness preceded by anorexia nervosa in identical twins. J. Clin. Psychiat. 45: 262–266, 1984.
47. Jones, D. J. Fox, M. M., Babigian, H. M., and Hutton, H. E. Epidemiology of anorexia nervosa in Monroe County, New York: 1960–1976. Psychosom. ed. 42:551–558, 1980.
48. Kanis, J. A., Brown, P., Fitzpatrick, K., Hibbert, D. J., Horn, D. B., Nairn, I. M., Shirling, D., Strong, J. A., and Walton H. J. Anorexia nervosa: a clinical, psychiatric, and laboratory study. Q. J. Med. 43:321–338, 1974.
49. Katz, J. L. Eating disorder and affective disorder: relatives or merely chance acquaintances? Compr. Psychiat. 28:220–228, 1987.
50. Kay, D. W., and Leigh, D. The natural history, treatment and prognosis of anorexia nervosa based on a study of 38 patients. J. Ment. Sci. 100:411–431, 1954.
51. Kendell, R. E., Hall, D. J., Hailey, A., and Babigian, H. M. The epidemiology of anorexia nervosa. Psychol. Med. 3:200–203, 1973.
52. King, A. Primary and secondary anorexia nervosa syndromes. Brit. J. Psychiat. 109:470, 1963.
53. Kron, L. Katz, J. L., Gorznski, G., and Wiener, H. Anorexia nervosa and gonadal dysgenesis. Arch. Gen. Psychiat. 34:332–335, 1977.
54. Larocca, F. E. F. Concurrence of Turner's syndrome, anorexia nervosa, and mood disorders: case report. J. Clin. Psychiat. 46:296–297, 1985.
55. Lasègue, E. C. De l'anorexia hystérique. Arch. Gen. Med. 21:385, 1873.
56. Morgan, H. G., and Russell, G. F. M. Value of family background and clinical features as predictors of long-term outcome in anorexia nervosa: four-year follow-up study of 41 patients. Psychol. Med. 5:355–371, 1975.

57. Pope, H. G., Jr., and Hudson, J. I.   Antidepressant drug therapy for bulimia: current status. J. Clin. Psychiat. 47:339–345, 1986.
58. Pops, M. A., and Schwabe, A. D.   Hypercarotenemia in anorexia nervosa. JAMA 205:533–534, 1968.
59. Rajs, J., Rajs, E., and Lundman, T.   Unexpected death in patients suffering from eating disorders. Acta Psychiat. Scand. 74:587–596, 1986.
60. Robinson, P. H., and Holden, N. L.   Bulimia nervosa in the male: a report of nine cases. Psychol. Med. 16:795–803, 1986.
61. Russell, G. F. M.   Metabolic aspects of anorexia nervosa. Proc. R. Soc. Med. 58:811–814, 1965.
62. Russell, G.   Bulimia nervosa: an ominous variant of anorexia nervosa. Psychol. Med. 9:429–448, 1979.
63. Schwabe, A. D., Lippe, B. M., Chang, R. J., Pops, M. A., and Yager, J. Anorexia nervosa. Ann. Intern. Med. 94:371–381, 1981.
64. Simmons, R. C., and Kessler, M. D.   Identical twins simultaneously concordant for anorexia nervosa. J. Am. Acad. Child Psychiat. 18:527–536, 1979.
65. Stunkard, A.   New Therapies for the eating disorders. Arch. Gen. Psychiat. 26:391–398, 1972.
66. Swift, W. J.   The long-term outcome of early onset anorexia nervosa. J. Am. Acad. Child Psychiat. 21:38–46, 1982.
67. Szmukler, G., McCance, C., McCrone, L., and Hunter, D.   Anorexia nervosa: a psychiatric case register study from Aberdeen. Psychol. Med. 16:49–58, 1986.
68. Theander, S.   Anorexia nervosa: a psychiatric investigation of 94 female patients. Acta. Psychiat. Scand. Suppl. 214:5–194, 1970.
69. Thomas, J. P., and Szmukler, G. I.   Anorexia nervosa in patients with Afro-Caribbean extraction. Brit. J. Psychiat. 146:653–656, 1985.
70. Tolstrup, K., Brinch, M., Isager, T., Nielsen, S., Nystrup, J., Severin, B., and Olesen, N. S.   Long-term outcome of 151 cases of anorexia nervosa. Acta Psychiat. Scand. 71:380–387, 1985.
71. Toner, B. B., Garfinkel, P. E., and Garner, D. M.   Long-term follow-up of anorexia nervosa. Psychosom. Med. 48:520–529, 1986.
72. Vandereycken, W., and Van den Broucke, S.   Anorexia nervosa in males. Acta Psychiat. Scand. 70:447–454, 1984.
73. Wålinder, J., and Mellbin, G.   Karyotyping of women with anorexia nervosa. Brit. J. Psychiat. 130:48–49, 1977.
74. Warren, W.   A study of anorexia nervosa in young girls. J. Child Psychol. Psychiat. 9:27–40, 1968.
75. Williams, J., King, M.   The "epidemic" of anorexia nervosa: another medical myth? Lancet I:205–207, 1987.

12. The Psychiatric Examination

Such is man that if he has the name for something, it ceases to be a riddle.

ISAAC BASHEVIS SINGER

The purpose of a psychiatric examination is to evaluate psychological function and to diagnose psychiatric disorders. To elicit enough information about the disorders in order to diagnose them, one must know their signs and symptoms, course and complications. That is why this chapter ends the book.

There is an art to eliciting clinical information. It can be learned in a formal way, but only in part. To establish the trust and rapport between doctor and patient that brings out reliable information empathy, intuition, and common sense are essential. These cannot be learned from books. Here we will give advice on interviewing, provide a logical framework for organizing observations, and suggest how case histories should be presented. But first a few words about terminology and time.

A *mental status* examination is the part of the physical examination that deals with the patient's thoughts, feelings, and behavior at a particular point in time. This term is often used as a synonym for a *psychiatric examination*, but *mental status* refers to one part of the psychiatric examination: the current thoughts, feelings, and behavior of the patient. *Psychiatric examination* includes the past history of the patient as well. The distinction between the two terms is somewhat artificial. As with *liver status* or *cardiac status*, what exists now is inseparable from what came before; a certain amount of historical background is unavoidable in describing the current mental status of a patient. Still, the term "mental status" is used when the *primary* focus of the questioning is on current functioning.

Internists, family practitioners, and other nonpsychiatrists generally have little time to conduct a physical examination: often no more than fifteen or twenty minutes. If the examination is "complete," it will include some attention to the mental status of the patient. This may be limited to a few minutes.

Later we will provide some screening questions that will help nonpsychiatrists decide quickly whether the mental state of the patient is abnormal. For now, we will assume the mental status examination is being conducted by a psychiatrist, a student or a resident in psychiatry who has the luxury of being able to spend a fair amount of time with the patient, observing him or her and asking questions.

Advice on Interviewing

Here are four rules for conducting a psychiatric examination.

1. *Start open ended.* Unless the patient is uncooperative or incapable of free expression (perhaps because of physical disability), let the patient tell his or her story with little or no interruption during the first five to fifteen minutes. After exchanging friendly greetings with the patient and attempting to set a relaxed tone, the interviewer should ask an open-ended question such as, "What is the problem that brings you here?" "What can I do for you?"

The patient is often tense early in the interview and this tension may indeed stimulate the information flow. A patient with a *formal thought disorder* (where the thoughts do not connect coherently) will reveal this quickly. Much of the information needed for the diagnosis is often provided in the first few minutes if the patient proceeds uninterruptedly. By steering the questions along certain lines, one may miss important material.

On the other hand, for a particularly tense patient, more structure at the beginning of the interview may lead to easier communication. With such a patient, the interviewer can ask specific questions that are emotionally neutral. Questions about the patient's background—where the patient grew up and went to school, marital status, job, other physicians seen—usually are not difficult to answer and provide a comfortable transition into asking about the presenting problems.

2. *Ask specific questions later.* One purpose of the mental status examination is to make a diagnosis, if possible. This requires specific questions if merely to rule out remote possibilities. For example, patients often avoid volunteering information about hallucinations. "Do you hear voices or see things that others do not hear or see?" or some variation on this query is often necessary to determine whether the patient is psychotic. "Do you feel in danger?" may elicit persecutory delusions. "Do you have a special mission in life?" may bring out grandiose delusions. "What are your plans after leaving the hospital?" may bring out unrealistic thinking, raising questions about judgment.

Even with the advantage of a long interview, the psychiatric examiner must ask specific questions bearing on a reasonable differential diagnosis. There is usually no point, for example, in going through a complete review of systems if the patient experiences excellent health and presents with symptoms of a psychiatric condition in which physical symptoms do not usually play a role.

Details about early life experiences rarely bear on the problem of making a differential diagnosis in adults. School and social history are often important but not always, particularly in dealing with elderly people.

3. *Establish the chronology of the illness.* Kraepelin (3) noted that the course of a psychiatric illness is as important as the symptoms. Sydenham (4) said that "true" illnesses should have common symptoms and a common course. Few, if any, pathognomonic symptoms exist in psychiatry. We agree with Kraepelin that establishing the course of an illness is as important as recognizing current symptoms.

When did the symptoms begin? Was the patient ever free of psychiatric symptoms? When? Age of onset is an important clue to diagnosis as many conditions typically begin at particular times in life. Has the illness been continual, always present with fluctuations, or episodic in the sense that symptoms sometimes go away entirely? How rapid was the onset? (Psychotic illnesses with abrupt onsets generally have a better prognosis than those with a gradual onset.) Have professional interventions (medications, psychotherapy) altered the course of illness? In general, has the patient tended to improve or get worse?

"Diagnosis is prognosis" is an old saying in medicine, and knowledge of the course of illness as well as the symptoms forms the basis for determining prognosis.

4. *Be friendly, sympathetic, respectful.* Examiners should never insult patients. They should never make fun of them. This may seem obvious, but there are subtle ways of betraying disrespect. Adult patients should be called "Mr." or "Mrs." or "Ms.," at least until the examiner knows them well.

Be sensitive to the emotional state of the patient. If certain questioning makes the patient angry, anxious, depressed, or tearful, this may offer an opportunity to enhance the patient's ability to communicate, though sometimes a return to more neutral ground is indicated so that the patient is not overwhelmed by the emotion.

A word about the uncooperative patient: To say, "I can't help you unless you help me" sometimes works, but usually it does not. Asking specific questions such as "What led to your coming here?" or "Whose idea was it that you come here?" may help lower resistance. Sometimes the interview must be postponed until another time when the patient may be more helpful. Anger toward the uncooperative patient is never appropriate.

Psychiatry, probably more than any other specialty, benefits greatly from informants—friends and family who will tell what the patient will not (or cannot). Although caution should be exercised in judging the merit of such information, it can be very helpful in making a diagnosis.

The Decision Tree

Except for open-ended questions at the beginning and specific questions toward the end, history taking should flow easily and casually, as in a conversation. Patients should be permitted to talk about what they want to talk about, but they should be gently guided back into channels that provide information the examiner requires for a diagnosis. From the minute a patient walks into the examination room, however, the examiner's mental "computer" starts making decisions. How is the patient dressed and groomed? Does the patient have a normal gait and range of motion? Is the patient hostile or friendly? How old does the patient appear to be?

Based on these first impressions, the interviewer starts narrowing the diagnostic range. The examiner's choices about *probable* diagnoses will determine which areas to emphasize and which to skip over lightly or omit entirely. The examiner's mind, indeed, functions as a computer. By the end of the interview—if it is successful—the choices will have narrowed to one or a few.

Figure 12.1 shows a highly simplified branching process for approaching the diagnosis of psychiatric disorders. The first decision concerns *memory*. If the patient has a normal memory, move to the right of the line in Figure 12.1. The second decision concerns psychosis. Is the patient psychotic or nonpsychotic? Psychosis can be both broadly or narrowly defined. Broadly defined, it refers to the gravity or seriousness of the condition, so that a suicidal patient might be called psychotic because suicide is serious. Narrowly defined, as here, psychosis means the presence of persistent hallucinations and/or delusions and/or disordered thoughts.

As shown in Figure 12.1, a *psychotic* person with a *normal memory* may suffer from schizophrenia, an affective disorder, or drug intoxication. Hallucinogens, amphetamines, and phencyclidine (PCP) are commonly associated with psychosis in the presence of a normal memory.

If the patient has a normal memory and is not psychotic, diag-

FIGURE 12.1 Major branches in diagnosis making.

Impaired Memory		Normal Memory	
Acute (Delirium)	Chronic	Psychotic	Nonpsychotic
	Dementia Retardation	Schizophrenia	Anxiety Disorders
		Acute	Panic
		Chronic	Obsessional
		Affective Disorders	Phobic
		Depression	Somatization Disorder
		Mania	Antisocial Personality
		Both	Chemical Dependence
		Drugs	Affective Disorders
			"Personality Disorders"

nostic possibilities include the anxiety disorders (there are eight in DSM-III-R), somatization disorder (hysteria), antisocial personality, drug (chemical) dependence, affective disorders, and other personality disorders. Thus, in some conditions such as affective disorders and drug dependence, the patient may or may not be psychotic. The term "other personality disorders" is included for purposes of completeness, but these conditions are still either too vaguely defined or too poorly studied to be useful diagnostic categories.

In most hospitals, about one-fifth of the patients who clearly have psychiatric abnormalities do not fit any of the categories in Figure 12.1. The suitable label for these people is *undiagnosed*. One advantage of this term is that physicians who deal with the patient in the future will not be biased by having a poorly grounded diagnosis in the chart. Another advantage is the sense of modesty it correctly implies.

A disadvantage is that many insurance companies *require* a diagnosis. In this case, one can include the most likely diagnosis or diagnoses prefaced by the term "rule out." Most insurance companies will accept this practice.

Outside of hospitals, many patients who consult psychiatrists do not have a diagnosable illness. They even lack symptoms of sufficient severity to justify being called "undiagnosed." One diagnosis for these more-or-less normal people who see psychiatrists is "problem of living," which suggests, if nothing else, that no conventional diagnosis seems to fit them.

Left of the center line in Figure 12.1 are conditions associated with impaired memory. Acute refers to those of recent onset (less than a month); chronic to those of longer duration. "Organic brain disorder" or "organic brain syndrome" are commonly used terms for patients with impaired memories. An acute brain disorder includes delirium. Dementia is a chronic brain disorder that represents a deterioration from a previous level of normal cognitive function.

As noted earlier, the first and in some ways most important decision concerns whether the patient has an impaired or normal memory. Although IQ tests measure more than memory, performance depends to a large extent on what persons have learned and how well they can recall it, that is, memory. Though memory loss may affect some functions more than others, *gross* memory disturbance usually affects intellectual functioning across the board.

To be "impaired" the patient's memory must be *really* impaired, in contradistinction to "normal forgetting." (Sometimes, admittedly, the distinction is not easy to make.) Disorientation is a form of memory impairment. Inability to do simple arithmetic reflects a bad memory, assuming the person once knew how to do arithmetic.

Diagnosing dementia is best done by estimating what the patient *should* know. If patients are interested in sports, they should be able to name sports figures. If interested in gardening, they should be able to name plants. Patients are often asked to do "serial sevens," whereby they subtract seven from one hundred and then continue subtracting sevens in a descending scale. In fact, normal people with little talent in arithmetic have trouble with serial sevens and their failure to do them may not be clinically significant. (Serial sevens tests attention and concentration as well as memory.) On the other hand, if certified public accountants cannot do serial sevens, dementia is likely, although their performance may have faltered because they were anxious or distracted.

Delirium is usually accompanied by agitation and autonomic hyperactivity as well as hallucinations and delusions (perhaps more often illusions: misinterpretation of stimuli). Sometimes delirious patients lie quietly in bed but still misidentify people and cannot remember the year or where they are. Delirious people may be dangerous. To escape their delusional persecutors they may jump out of hospital windows or attack those around them. They must be watched closely.

Poor intellectual functioning is associated with, and often indistinguishable from, bad memory. Impaired memory *produces* impaired intellectual functioning. As noted, "intelligence" encompasses more than memory, but even those skills not normally associated with memory (e.g., reasoning ability) often suffer when memory is impaired.

Depressed patients sometimes have bad memories and this is called "pseudodementia." Their memory improves as their depression improves. Stroke patients and patients with Alzheimer's disease also experience depression, but *their* memory usually does not improve as the depression improves.

Here is the important point about gross memory impairment (as distinguished from absentmindedness, normal forgetting, or "not paying attention"): Patients with organic brain disorders may display *any* psychiatric symptom associated with disorders on the right side of Figure 12.1, *but organic brain disorders still take precedence as the diagnosis* unless the other disorders clearly antedated the brain disorder. Anxiety, depression, delusions, hallucinations, mania, incoherence, obsessions, phobias may all occur in organic brain disorders. Diagnosing organic brain disorders is one of the most important things a psychiatrist can do. It initiates a search for the *cause* of the disorder and the cause may be *treatable*. Physicians are uniquely qualified to identify the organic disease. Knowing anatomy, physiology, and biochemistry and aided by modern imaging and laboratory techniques, physicians can evaluate the entire range of causes of organic brain disorders, including brain tumor, endocrine and metabolic disorders, and infections.

The Mental Status Format

The purpose of the mental status format is to help the interviewer organize and communicate his or her observations about a patient. Minor deviations occur in the format from expert to expert, but *some* framework for observations is necessary to facilitate thinking and communication. The format presented here is commonly used and includes the following categories:

Appearance and behavior
Form and content of thought

Affect and mood
Memory and intellectual functioning
Insight and judgment

Appearance and Behavior

The patient's appearance is often relevant to the diagnosis. Schizophrenics, for example, are often poorly groomed and sometimes dirty. Depressives also are often negligent about their dress and grooming. A manic may wear a funny hat. Sunglasses worn indoors may suggest paranoia; tattoos often suggest antisocial personality; a puffy face and red palms are suggestive of, but not diagnostic of, alcoholism.

If the patients look older than their stated age, this may suggest depression or long-term substance abuse. If this is the case, they may begin to look younger as they recover.

The patient's attitude toward the interviewer may be significant. Paranoids are often suspicious, guarded, or hostile. Hysterics sometimes try to flatter interviewers by comparing them favorably with previous doctors; they are often dramatic, friendly—sometimes seductive. Manics may crack jokes and occasionally are quite funny—when they are not irritable or obnoxious. Sociopaths may seem like con men—and sometimes are.

The patient may be agitated—unable to sit still, moving constantly. Others are retarded, slumping in their seats, slow in movement and speech. Talking may seem an effort.

Agitation and retardation can have several causes. Neuroleptic drugs may produce a restlessness called "akathisia," in which the patient cannot sit still and feels compelled to walk. Neuroleptics also may produce Parkinson-type symptoms, including tremor and an expressionless face. Pacing and handwringing may be expressions of depression; joviality and volubility of mania.

Neuroleptic drugs are given so commonly that it is often impossible to determine whether abnormal movements are drug induced or catatonic symptoms. In fact, similar involuntary movements were observed in schizophrenics years ago before drugs were introduced. It is said that catatonic symptoms are disappearing, but what previously was called catatonic may now be interpreted as drug induced without knowing whether drugs are responsible or not.

Schizophrenia also may involve psychomotor disturbances such as mannerisms, posturing, stereotypical movements, and negativism (doing the opposite of what is requested). Also seen is *echopraxia*, in which movements of another person are imitated, and *waxy flexibility*, in which awkward positions are maintained for long periods without apparent discomfort. Some patients say nothing. This is called "mutism"; it may be seen in schizophrenia, depression, brain syndrome, and drug intoxication.

Form and Content of Thought

Form refers to intelligibility related to associations: Does the patient have "loose associations" in the sense of being circumstantial, tangential, or incoherent? Elderly people are often circumstantial. They return to the subject but only after providing excessive detail. Tangentiality is a flow of thought directed away from the subject being inquired about, with no return to the point of departure. Schizophrenics are often tangential. Pressure of speech and flight of ideas are seen in mania and in drug intoxication. With pressure of speech, the patient seems compelled to talk. Manic speech flits from idea to idea, sometimes linked by only the most tenuous connections. Unlike tangentiality, however, manic speech frequently has connections that can be surmised. Manics often rhyme or pun and make "clang" associations, using one word after another because they sound similar. Manics tend to be overinclusive, including irrelevant and extraneous details.

Derailment, often seen in schizophrenia, is a form of speech in which it is impossible to follow the logic of the associations. Sometimes schizophrenics invent new words (neologisms) that presumably have a private meaning. Sometimes schizophrenics display poverty of thought, conveying little information with their words. *Echolalia* refers to occasions when the patient repeats words back to the interviewer. Other abnormal speech patterns associated with schizophrenia (as well as dementia) are perseveration, in which the patient seems incapable of changing topics, and blocking, in which the flow of thought is suddenly stopped, often followed by a new and unrelated thought.

When patients persistently display any of these symptoms (excluding poverty of thought), they are said to have a *formal thought disorder*, meaning that the structure or form of thinking is disordered.

Content of thought refers to what the patient thinks and talks about. Under this category come hallucinations, delusions, obsessions, compulsions, phobias and preoccupations deemed relevant to the psychiatric problem.

Delusions are fixed false ideas neither unamenable to logic social pressure nor congruent with the patient's culture. They should be distinguished from *overvalued* ideas in which a patient has a fixed notion that most people consider false but that is not entirely unreasonable or cannot be disproven, such as certain superstitions. Delusions occur in organic brain disorders, schizophrenia, affective disorders, and various intoxications.

Jaspers (2) believed the *subject* of the delusion had diagnostic significance. If the delusional ideas were "understandable," they more likely occurred in depressed patients. Understandable delusions included those in which persons were convinced they had a serious life-threatening illness such as cancer, were impoverished, or were being persecuted because they were bad persons. Jaspers pointed out that healthy, prosperous, and likable people often worry about their health, finances, and approval by others. Such delusions are thus understandable.

Delusions that are *not* understandable are seen in schizophrenia, according to Jaspers. Schizophrenic delusions tend to be bizarre; for example, one's acts are controlled by outside forces (delusions of control or influence) or one believes that one is Jesus or Napoleon. Schizophrenialike delusions occur often in amphetamine psychosis and, less commonly, in other intoxicated states (e.g., from cocaine or cannabis). The delusions of schizophrenia fall outside the ordinary person's experience: The examiner finds it difficult to identify with the schizophrenic's private world, hence the term "autism," derived from auto, is often applied to schizophrenic thinking.

Religious delusions are sometimes hard to interpret. Religious beliefs often seem delusional to those who do not accept the beliefs but normal to those who do. Among fundamentalist religious people, truly pathological delusions are usually identified without difficulty by others in the congregation.

Content also encompasses perceptual disturbances. In *illusions*, real stimuli are mistaken for something else (a belt for a snake). *Hallucinations* are perceptions without an external stimulus. *Auditory hallucinations* may consist of voices or noises. They are associated primarily with schizophrenia but occur in other conditions such as chronic alcoholic hallucinosis and affective disorders (1). *Visual hallucinations* are most characteristic of organic brain disorders, especially delirious states. They also occur with psychedelic drug use and in schizophrenia. (Certain hallucinations are more common in some conditions, but no type of hallucination is found exclusively in any illness.) *Hypnagogic hallucinations* arise in the period between sleep and wakefulness, especially when falling asleep. Their occurrence is normal except when they are a symptom of narcolepsy.

Olfactory hallucinations are sometimes associated with complex partial seizures that involve the temporal lobes. *Haptic* (tactile) *hallucinations* occur in schizophrenia and also in cocaine intoxication and delirium tremens. The sensation of insects crawling in or under one's skin (formication) is particularly common in cocaine intoxication, but it also happens in delirium tremens.

In *extracampine hallucinations*, the patient sees objects outside the sensory field (e.g., behind his head). In *autoscopic hallucinations*, the patient visualizes himself projected into space. The latter sensation is also known as *Doppelgänger* (seeing one's double).

Other perceptual distortions include *depersonalization* (the feeling that one has changed in some bizarre way), *derealization* (the feeling that the environment has changed), and déjà vu (a sense of familiarity with a new perception).

In one study of nonpsychiatric patients, 40 percent reported hallucinations, particularly seeing dead relatives. They had no other psychiatric symptoms and the hallucinations were not judged to be clinically important. Thus a history of transient hallucinations or other perceptual disturbances, which occur occasionally during exhaustion or grief, does not necessarily signify the presence of psychosis. They must be interpreted in the context of the overall clinical picture.

Affect and Mood

Affect refers to a patient's outwardly (externally) expressed emotion, which may or may not be appropriate to her reported mood and content of thought. For example, if a person smiles happily while telling of people trying to poison her, the affect would be described as inappropriate. If one describes unbearable pain but looks as if she were discussing the weather, the affect again would be inappropriate.

Affect is sometimes referred to as "flat," meaning that the usual fine modulation in facial expression is absent. Schizophrenics sometimes have a flat affect, but so do patients taking neuroleptic drugs, and a depressed patient may show little change of expression while speaking.

"Flat affect" is probably the most overused and misused term in the psychiatric examination. It should only be used if the affect is extremely "flat" or "blunted." Inappropriate and flat affects are especially associated with schizophrenia.

Sometimes hysterics have an inappropriate affect in that they describe excruciating pain and other extreme distress with the same indifference or good cheer with which they would describe a morning of shopping. (The French call this "la belle indifférence.")

Mood refers to what the patient says about his internal emotional state. "I am sad," "I am happy," "I am angry" are examples. Mood and affect are sometimes *labile*, meaning that there is rapid fluctuation between manifestations of happiness, sadness, anger, and so on. Labile affect is often seen in patients with organic brain disorders.

Memory and Intellectual Functioning

Subsumed under memory is orientation, meaning orientation for person, place, and time. To be disoriented for time, the patient should be more than one day off the correct day of the week and more than several days off the current date. Misidentifying people (thinking the nurse is one's aunt) is a clear case of disorientation, as is giving the wrong year or the wrong city and wrong hospital where one is currently residing. This part of the mental status is exceedingly important because, if a patient has a gross memory impairment (and is not malingering), he or she almost always has an organic brain disorder and all other psychiatric symptoms may be explainable in this context. (The pseudodementia of depression is one exception.)

There are many tests for memory and intellectual functioning. Memory can be subdivided into immediate, short-term, recent and remote memory. Serially subtracting seven from one hundred is a test of immediate memory (assuming the person's arithmetic was ever adequate for the task) as well as a test of attention and concentration. Short-term-memory loss can be tested by asking patients to remember three easy words you have spoken or by showing them three objects and then, five to fifteen minutes later, asking them to repeat what they heard or saw. A short-term memory deficit is the sine qua non of Korsakoff's syndrome. Recent memory refers to recall of events occurring in recent days, weeks, or months; remote memory involves recall of events occurring many years before, such as the winner of a long-ago presidential election. In dementia, recent memory is usually more severely impaired than remote memory.

As noted earlier, tests of intellectual functioning should be interpreted with the patient's background, education, cooperativeness and mood state in mind. A history major should be able to name seven presidents, but a "normal" person with a third-grade education may not be able to do so. Depressed patients may be too slowed down or distractible to concentrate. One approach is to ask patients about their interests and then test their fund of information in those areas.

Insight and Judgment

A person who has insight will know whether he is (or was) psychiatrically ill. If he says, for example, that the voices are "real," he lacks insight. If he says it was simply his imagination playing tricks on him, he has insight. If he says there is nothing wrong with him but that his evil uncle has arranged for his hospitalization because of a Communist conspiracy, he may or may not have insight. (Even paranoids, as the saying goes, sometimes have real enemies.) Psychosis and organic brain disorders are both associated with lack of insight; so-called neurotics usually realize they have something wrong with them.

The term "judgment" is used here in the same sense as "competence" is used in civil court: A competent person is able to understand the nature of the charges and to cooperate with counsel. It implies that a person is realistic about his limitations and life circumstances. A good question to ask is, "What are your plans when you leave the hospital?" If the patient says that he plans to start a chain of restaurants and has no money, this displays impaired judgment. Severe impairment of judgment is seen most often in dementia and psychotic disorders.

Excluding Psychiatric Disorders

Sometimes for all physicians and often for nonpsychiatric physicians, examination of the "mind" must be accomplished quickly, lest the liver, lungs, heart, and deep tendon reflexes be slighted. For the dozen disorders described in this book, a single question may suffice to exclude the possibility the patient has a given disorder. Some disorders will be missed, but one or two questions will identify the great majority of patients who do *not* have a particular psychiatric illness:

Depression: Ask if the patient sleeps well. If she sleeps well without medication, the chances of a serious depression are slight. (Oversleeping represents "not sleeping well" as much as undersleeping.)

Mania: Ask if the patient has ever gone on a spending spree. Most manics have, even manics who cannot afford it.

Schizophrenia: Ask if the patient has ever heard or seen things that other people did not hear or see. Ask if he has ever been afraid of being poisoned or controlled by external forces. Hallucinations sometimes occur in normal people, but the presence of both hallucinations and delusions in a person with more or less normal mood suggests schizophrenia.

Panic disorder (anxiety neurosis): Has the patient ever thought she was having a heart attack that did not occur? Does she ever become intensely apprehensive for no apparent reason? Anxiety neurotics report both. At church, does she find a seat on the aisle close to the back? Anxiety neurotics almost always do. They feel the need to make a quick exit if a panic attack seems impending.

Hysteria (somatization disorder): Hysterics are mostly women. Get a menstrual history. If the woman denies having problems with her menstrual periods—if she has never missed work or school because of dysmenorrhea—hysteria is unlikely. If she has reached the age of thirty-five without having her appendix removed, plus some other elective operation, she is probably not hysteric.

Obsessive compulsive disorder: Does the patient, sitting in a waiting room, count things, such as the number of tiles on the floor? Does the patient repeatedly check a door to see if it is locked or an oven to see if it is turned off? Counting and checking are so common in this disorder that, if absent, the diagnosis should be questioned.

Phobic disorders: Does the patient avoid certain situations because they frighten him? Is the fear unreasonable?

Alcoholism: Has the patient ever stopped drinking for a period of time? If so, and the reason is not medical or a desire to lose weight, the patient probably stopped because he was worried about his drinking. At this point, the clinician can ask *why* he was worried, and this may break down the denial that is characteristic of alcoholism. Almost every alcoholic has stopped, or tried to stop, at some time in his life. This is a better approach than asking, "Do you drink too much?"

Drug dependence: "Have you ever worried about a drug habit?" is probably as good an opener as any.

Antisocial personality (*sociopathy*): Ask if the patient was frequently truant in grade and high school. Rare is the sociopath who did not cut classes and get in trouble with school authorities as a teenager.

Dementia: Ask if the patient forgets where she parks her car. If this happens often, there should be some concern about her memory. Or simply ask, "How is your memory?" Many people with memory problems are relieved to have the chance to talk about them.

Anorexia nervosa: If the person is intelligent, ask her (and it is usually a her) if she has ever been told she had anorexia nervosa. Anorectics usually know their diagnosis; the press is full of it. Does the patient stuff herself (or himself) with food and then induce vomiting. This practice, called "bulimia," often goes with anorexia in both sexes.

Another question: "Are you the right weight?" If the patient is 5 foot 7 inches, weighs 92 pounds, is not a model, and says "I'm too fat," the diagnosis is made.

Sexual problems: "Do you have a sex problem?" is usually sufficient. Since the Sexual Revolution, people are not as reticent about sexual matters as they once were.

These questions when answered in the negative will eliminate most people who have the above disorders. There will be few false negatives. There will be many false positives. (Many people sleep poorly and sit at the back of churches who do not have depression or anxiety disorders.) But for the physician trying to rule out disorders, false positives are unimportant. They simply mean probing is required. Probing takes time and referral to a psychiatrist may be in order.

Suggestions for Presenting Cases

There is obviously a good deal of latitude in presenting case histories for teaching purposes. Different institutions and different teachers within the institutions will have their own advice on the subject. However, discussions with these teachers reveal some agreement about certain points. Here are some general rules for presenting patients.

1. Don't read the history.

2. Don't exceed ten to fifteen minutes (allowing for interruptions).

3. Start with *identifying data*: name, age, race, marital status, vocation.

4. Provide a clue to the problem you will highlight, e.g., "This patient presents a diagnostic problem," "He has not responded to standard treatments," "She comes from an unusual family." Such clues offer a framework for your audience into which the rest of the presentation will fit.

5. Avoid dates. Open with "Patient was admitted to [hospital] _____ (days, weeks, months) ago. Do not refer to events occurring on December 3, 1937, but say, "At the age of 15, the patient _____." Instead of saying "Between November and January of 1955 and 1956," say "For a three-month period when the patient was twenty years old, he _____." It may be easier for patients to remember events by dates, but the listener has to translate dates into ages and, for the unmathematically inclined, this may be difficult while concentrating on the presentation.

6. Begin with the *psychiatric history*. A good way to begin is, "The patient had no psychiatric problems until age _____ (or _____ days, weeks, or months ago) when he (slowly or rapidly) developed the following symptoms _____"; then list

the symptoms in order of severity. Tell how long the symptoms persisted (for weeks, months, years, or to the present) and what happened as a result (hospitalization, other treatment, full or partial recovery).

Often, of course, establishing time of onset is difficult or impossible, particularly when dealing with a poor historian or a complicated case. The onset of illness in a mentally retarded person would be "from birth," which does not help much. But an *attempt* to establish onset can be of considerable help because different illnesses characteristically begin at different ages.

7. It is important to know whether the illness has been *chronic*, perhaps with fluctuations, or *episodic* with full remissions between episodes. If the patient has had more than one episode, describe subsequent episodes, briefly giving the same information that was given for the first episode. Symptoms and life events obviously are interrelated, but emphasize the symptoms rather than the life events unless the life events appear to be causally related to the symptoms.

8. A brief *family history* should include the following: Whether a close blood relative of the patient had a serious psychiatric illness requiring treatment (and what the treatment was, if known), pertinent medical illnesses, and suicide, alcohol or drug problems.

9. *Social history* should include (very briefly) circumstances of upbringing, particularly whether the parents were divorced or separated or whether the patient was brought up by both parents; parental vocation; siblings; years of education and how well the patient did in school from the standpoint of grades and adjustment; military and job history; marital history; and number and ages of children.

10. Review the *medical history* only as it is pertinent to the psychiatric problems. The same applies to the review of systems, physical findings, and laboratory results.

11. Give the *mental status* as it was obtained either on admission or at the first opportunity to fully examine the patient. The mental status findings should be presented in the order provided in the previous section.

12. End the presentation with *course in hospital*. Tell how the

patient has been doing, whether he has improved, what treatment he is receiving. In other words, bring the patient up to the present moment.

13. With rare exceptions, all this can be presented in ten to fifteen minutes. The trick is to keep in mind at all times the goal of the presentation. If it is diagnostic, the differential diagnosis and the points for and against each of the reasonably likely diagnoses should be given. If you start out by saying the patient was psychiatrically well until the age of sixty, dwelling on such diagnoses as mental retardation, schizophrenia, somatization disorder, or panic disorder is unlikely to be useful. Assuming the history is correct (though, granted, this is often a dubious assumption), people who are well until the age of sixty and then develop major psychiatric problems generally have either an affective disorder or an organic brain syndrome.

14. The reasons for presenting the history and mental status according to the above sequence is to avoid leaving out important information and to make it easier for the listeners to follow the narration. There are many variations on this format, none perfect (people's lives are much more complicated than formats). Unlike written psychiatric histories, however, oral presentations should not attempt to be comprehensive. They should touch on the following categories, but not all with equal emphasis.

HISTORY
Identifying data
Focus of the presentation
Psychiatric history
Family history
Social history
Medical history
Review of systems
Physical findings
Laboratory results

MENTAL STATUS
Appearance and behavior
Form and content of thought

Affect and mood
Memory and intellectual functioning
Insight and judgment

References

1. Goodwin, D. W., Alderson, P., and Rosenthal, R.   Clinical significance
   of hallucinations in psychiatric disorders. Arch. Gen. Psychiat. 24:76–80,
   1971.
2. Jaspers, K.   *General Psychopathology*. Chicago: Univ. of Chicago Press,
   1963.
3. Kraepelin, E. Dementia Praecox and Paraphrenia (trans. R. M. Barclay;
   G. M. Robertson). Edinburgh: E. S. Livingstone, 1919.
4. *Selected Works of Thomas Sydenham, M.D.*   London: John Bales & Sons,
   Danielson, 1922.

Index